Springer Optimization and Its Applications

VOLUME 106

Aims and Scope
Optimization has been expanding in all directions at an astonishing rate during the last few decades. New algorithmic and theoretical techniques have been developed, the diffusion into other disciplines has proceeded at a rapid pace, and our knowledge of all aspects of the field has grown even more profound. At the same time, one of the most striking trends in optimization is the constantly increasing emphasis on the interdisciplinary nature of the field. Optimization has been a basic tool in all areas of applied mathematics, engineering, medicine, economics, and other sciences.

The series *Springer Optimization and Its Applications* publishes undergraduate and graduate textbooks, monographs and state-of-the-art expository work that focus on algorithms for solving optimization problems and also study applications involving such problems. Some of the topics covered include nonlinear optimization (convex and nonconvex), network flow problems, stochastic optimization, optimal control, discrete optimization, multi-objective programming, description of software packages, approximation techniques and heuristic approaches.

More information about this series at http://www.springer.com/series/7393

Antonio José Vázquez Álvarez
Richard Scott Erwin

An Introduction to Optimal Satellite Range Scheduling

 Springer

Antonio José Vázquez Álvarez
National Research Council
Albuquerque, NM, USA

Richard Scott Erwin
Air Force Research Laboratory
Space Vehicles Directorate, Kirtland
AFB, NM, USA

ISSN 1931-6828   ISSN 1931-6836   (electronic)
Springer Optimization and Its Applications
ISBN 978-3-319-25407-4   ISBN 978-3-319-25409-8   (eBook)
DOI 10.1007/978-3-319-25409-8

Library of Congress Control Number: 2015953220

Mathematics Subject Classification (2010): 68M20, 90B35, 90B36, 90-02, 68Q25, 90C39, 97K30, 05C85, 91A40, 91A80, 91A10, 91A18, 62C10

Springer Cham Heidelberg New York Dordrecht London

Printed on acid-free paper

Springer International Publishing AG Switzerland is part of Springer Science+Business Media (www.springer.com)

To my parents.   A. J.
To my wife, my son, and my parents.   R. S.

Preface

The problem of scheduling interactions among satellites and ground stations has been around for decades, and most of the literature tackling this problem has focused on approximate solutions. In this book we have tried to find the optimal solution to this problem and some of its variants and also to unify criteria and notation across the satellite range scheduling literature. To our knowledge, this is the first work that will accomplish both objectives.

We wrote this book as a result of a 2-year (2013–2015) postdoctoral fellowship at the Air Force Research Laboratory, funded by the National Research Council. The major results contained herein were published as a series of conference and journal papers during this period. This work, although based on these publications, considerably extends them by solving new problems and binding them together as a whole.

This endeavor has not been easy, with telecommunication and control engineering backgrounds for tackling a problem from operations research. We think however that this combination provides increased value to this book, and we have in fact tried to make it accessible to those readers that are facing this problem for the first time. This book is also aimed at satellite operations engineers and scheduling algorithm designers, as we here provide reference (optimal) solutions to this problem and some of its most important variants.

We are conscious that we are only scratching the surface on satellite range scheduling, but we have tried to provide a strong framework over which keep finding solutions to more complex problems in this field. We hope this book keeps up with the high standards of previous literature in this field, and more importantly, we hope this book to be useful to students and algorithm designers.

Albuquerque, NM, USA
Kirtland AFB, NM, USA
April 2015

Antonio José Vázquez Álvarez
Richard Scott Erwin

Acknowledgments

I would like to thank my parents, Anun and Javier, for all their encouragement, not only during this period but throughout all my life. This work would not be possible without them, which always have been and will be my example. I want to thank my brother Luis too for his encouragement and support in the distance. Very especially I would like to thank my fiancée Nelmy Jerez, meeting her has truly changed my life, and her encouragement and love motivated me to continue with this endeavor.

I would like to thank my research advisor R. Scott Erwin, coauthor of this book, for all his help and trust during this period. He has always pointed the research in the right direction if I diverged too much, and since the beginning he has always shared his knowledge and gave me several pieces of valuable advice. I would also like to thank all the people I have met in AFRL, COSMIAC, and UNM, especially Alonzo Vera, Carolin Früh, Craig Kief, David Alexander, Frederick Leve, Moriba Jah, Robert Terselic, and many others who have been very helpful and friendly.

I am very grateful to the National Research Council[1] for this opportunity and for all their support during this period, especially to Peggy Wilson, who has always answered my never-ending lists of questions.

I would like to thank Springer for their trust in this project since its inception and for providing the LaTeX monograph template. Finally I would like to thank Springer, Scitepress, and IEEE for granting us permission for recasting some materials from our previous work. We have generated all the diagrams with DIA and Inkscape and the simulation maps and graphs with MATLAB and did the typesetting in MiKTeX.

A. J.

I would like to thank first my wife, Kim, and my son, Ian; it is the time I have spent away from them, or the times I have been home but not present, that are the real price paid for accomplishments such as this.

---

[1]This research was performed while the author held a National Research Council Research Associateship Award at the Air Force Research Laboratory (AFRL).

I would like to thank my parents, Richard and Barbara, for giving me the opportunities in life that led to the successes I have enjoyed.

Finally, I'd like to thank my coauthor, Antonio, for his incredible efforts producing the results herein as well as in preparing this monograph.

R. S.

Contents

Part I   Introduction

1   Motivation 3
   1.1   Motivation 3
   1.2   Why Optimal Scheduling? 4
   1.3   Why this Book? 4
   1.4   Structure of the Book 5
   1.5   Main Contributions 7
   References 9

2   Scheduling Process 11
   2.1   Scheduling Process 11
   2.2   Scheduler Characteristics 15
   2.3   Satellite Range Scheduling Problems 15
   2.4   Issues Beyond the Scope of this Text 16
   References 17

Part II   Satellite Range Scheduling

3   The Satellite Range Scheduling Problem 21
   3.1   Problem Formulation 21
      3.1.1   Model for the Scenario 22
      3.1.2   Model for the Requests 23
      3.1.3   Problem Constraints 25
      3.1.4   Schedule Metrics 32
   3.2   Complexity of SRS 34
      3.2.1   Introduction to Complexity Theory 34
      3.2.2   Complexity of the SRS Problem 34

    3.3   General Scheduling Problems 37
          3.3.1   Problem Classification... 37
          3.3.2   Problem Reducibility 39
    3.4   Relating Satellite and General Scheduling Problems................. 40
          3.4.1   One Machine Problems... 40
          3.4.2   Several Identical Machines Problems.......................... 41
          3.4.3   Several Unrelated Machines Problems 42
    3.5   Summary 44
    References 46

4   **Optimal Satellite Range Scheduling** 49
    4.1   Scenario Model for Fixed Interval SRS 49
    4.2   Optimal Solution for Fixed Interval SRS 50
          4.2.1   Description of the Algorithm.................................. 51
          4.2.2   Optimality of the Solution and Complexity
                  of the Algorithm 54
    4.3   Extension of the Algorithm... 56
          4.3.1   Optimal Discretized Variable Slack SRS 56
          4.3.2   Optimal Fixed Interval SRS with Redundancy 57
    4.4   Remarks on the Complexity 59
          4.4.1   Greedy Earliest Deadline Algorithm........................... 59
          4.4.2   Greedy Maximum Priority Algorithm 59
          4.4.3   About the Topology of the Scenario 60
          4.4.4   About the Number of Passes 60
          4.4.5   About Partial Results 60
    4.5   Graph Generation Example... 61
    4.6   Simulations 66
          4.6.1   Simulation: Practical Case 67
          4.6.2   Simulation: Worst Case....................................... 68
          4.6.3   Simulation: Number of Passes 70
          4.6.4   Simulation: Partial Results 70
    4.7   Summary 71
    References 73

Part III   Variants of Satellite Range Scheduling

5   **Noncooperative Satellite Range Scheduling**............................. 77
    5.1   Scenario Model for the SRS Game................................... 78
    5.2   Elements of the SRS Game 79
          5.2.1   Players.. 79
          5.2.2   Sequential Decisions.. 80
          5.2.3   Actions 80
          5.2.4   Shared Information 80
          5.2.5   Payoffs 80

          5.2.6   Rationality 81
          5.2.7   Extensive Form 81
     5.3 SRS Game with Perfect Information 82
          5.3.1   Description of the Algorithm................................ 83
          5.3.2   Stackelberg Equilibrium Solution............................ 87
          5.3.3   Computational Complexity.................................... 89
     5.4 Limited Information Versions of the Problem 89
          5.4.1   SRS Game with Uncertain Priorities.......................... 90
          5.4.2   SRS Game with Uncertain Passes 92
     5.5 Remarks on the SRS Game 93
          5.5.1   Equilibrium vs. Security Strategy 93
          5.5.2   Stackelberg Equilibrium vs. Nash Equilibrium 93
          5.5.3   Social Welfare and Price of Anarchy 94
          5.5.4   Machine Scheduling vs. SRS 94
     5.6 Graph Generation Example... 94
     5.7 Simulations 101
     5.8 Summary 104
     References 105

6 **Robust Satellite Range Scheduling**................................... 107
     6.1 Scenario Model for Robust SRS 108
          6.1.1   Complexity of the Robust SRS Problem....................... 110
     6.2 Restricted Robust SRS Problem 110
     6.3 Remarks regarding Multiple Scheduling Entities 114
     6.4 Variants of the Robust SRS Problem 116
          6.4.1   Robust SRS with Random Priorities.......................... 116
          6.4.2   Robust SRS with Random Durations 119
     6.5 Considerations on the Basic SRS Problem 123
     6.6 Schedule Computation Example 124
     6.7 Simulations 124
     6.8 Summary 126
     References 128

7 **Reactive Satellite Range Scheduling** 129
     7.1 Scenario Model for Reactive SRS................................... 129
          7.1.1   Complexity of the Reactive SRS Problem..................... 131
     7.2 Restricted Reactive SRS: Single Pass Update Model 132
          7.2.1   Overview of the Algorithm.................................. 132
          7.2.2   Preprocessing 133
          7.2.3   Recomputation 139
     7.3 Restricted Reactive SRS: Multiple Pass Update Model 140
     7.4 Schedule Computation Example 140
     7.5 Summary 145
     References 147

8   Summary..   149
   8.1   Conclusions   149
   8.2   Future Work   149

Glossary   153

Index..   159

Acronyms

AFSCN	Air Force Satellite Control Network
DAG	Directed acyclic graph
DSN	Deep Space Network
EOS	Earth observation satellite
ESA	European Space Agency
ESTRACK	ESA Tracking Station Network
FI	Fixed interval
Fig.	Figure
FNE	Fixed number of entities
GB	Gigabyte
GEO	Geostationary Earth orbit
GHz	Gigahertz
GS	General scheduling
LEO	Low Earth orbit
LOS	Line of sight
MuRRSP	Multiple resource range scheduling problem
N/A	Not applicable
NASA	National Aeronautics and Space Administration
NP	Nondeterministic polynomial
P	Polynomial
PoA	Price of anarchy
RAM	Random access memory
SiRRSP	Single resource range scheduling problem
SRS	Satellite Range Scheduling
SW	Social welfare
TVG	Time varying graph

Symbols

α	First term in unified notation
α_l	Failure probability for pass p_l
$\alpha_{j,m}$	Failure probability for pass $p_{j,m}$ in P_j^{w} associated to p_j^{w}
$\alpha_{j,k,m}$	Failure probability for pass $p_{j,k,m}$ in $P_{j,k}^{\mathrm{w}}$ associated to $\mathrm{j}_j^{\mathrm{w}}$
β	Second term in unified notation
$\beta_{j,m}$	Probability for the priority value $w_{j,m}$ for the pass p_j^{w}
$\beta'_{j,m}$	Probability for the duration value $\rho_{j,m}$ for the request $\mathrm{j}_j^{\mathrm{w}}$
$\beta'_{j,k,m}$	Probability for the pass $p_{j,k,m}$ in $P_{j,k}^{\mathrm{w}}$ associated to $\mathrm{j}_j^{\mathrm{w}}$
γ	Third term in unified notation
Γ_j	Finite discrete random variable for the durations of the pass p_j^{w}
Δt	Discretization time step
ρ_j	Duration of request j_j
$\overline{\rho_j}$	Maximum duration of request j_j
$\underline{\rho_j}$	Minimum duration of request j_j
$\overline{\overline{\rho}}$	Maximum duration among all requests
τ_{c}	Minimum time between priority changes
τ_{e_j}	End time of visibility window o_j
τ_{s_j}	Start time of visibility window o_j
ϕ	Sign of event
$\phi(p_m, p_l)$	Function for checking conflict between passes p_l and p_m
$\phi_g(p_k)$	Ground station associated to pass p_k
$\phi_s(p_k)$	Satellite associated to pass p_k
ψ_l	Sub-tree associated to node n_l in extensive form representation
ω_a	Payoff reduction constant for unfeasible paths
a_k	Priority normalization factor
a_{p_l}	Action performed for pass p_l
A_{b}	Algorithm for generating the backward graph
A_{f}	Algorithm for generating the forward graph
$A_i(l)$	Nodes added to the frontier B_i for node n_l

$\mathbf{A}_i$	Set of actions for satellite s_i
A_p	Algorithm for finding best pairs in the overlaid graphs G_f and G_b
A_r	Algorithm for finding the alternative longest paths
b	Edge in backward graph
B_i	Frontier in the graph
C_Σ	Unitary capacity (unified notation)
$C_\Sigma(P')$	Total number of conflicts for the schedule P'
$C_G(P')$	Number of ground station conflicts for the schedule P'
$C_S(P')$	Number of satellite conflicts for the schedule P'
C_x	m-ary capacity (unified notation)
d	Number of passes per day
d_j	Due time of request j_j
$\eth_j$	Due time of job $\mathfrak{J}_j$ (unified notation)
$\mathfrak{D}$	Multiple unrelated machines, distributed scheduler (unified notation)
$\mathscr{D}$	Dismissing pass action
$D_i(l)$	Nodes deleted from the frontier B_i for node n_l
D_l	Set of later nonconflicting passes for p_l
D_n	Transformation of requests into passes
e, e^-, e^+	Event, end time event, and start time event
E, E^-, E^+	Set of events, set of end time events, and set of start time events
$f_j^w(t)$	Priority function for request j_j
$f_e^-(p_l)$	Function for generating end time event from pass p_l
$f_e^+(p_l)$	Function for generating start time event from pass p_l
g_i	Ground station i
G	Set of ground stations
G_b	Backward graph
G_f	Forward graph
H_l	History of play until pass p_l
$H_{i,l}^s$	History of play for satellite s_i until pass p_l
$I_i(l)$	Information available to player i at stage l
I_{PI}	Perfect information
I_{UP}	Uncertain passes information
I_{UW}	Uncertain priorities information
j_j	Request j
$j(e_i)$	Request associated to pass associated to event e_i
J	Set of requests
J_{FS}	Set of requests with fixed slack
$\mathfrak{J}_i$	Job i (unified notation)
$J_i(P')$	Payoff for satellite s_i and schedule P'
J_j^a	Set of requests that are active for node n_j
J_{NS}	Set of requests with no slack
J_{VS}	Set of requests with variable slack
J^w	Set of requests with random durations

$J(n_j)$	Set of requests associated to backtracking from n_j to n_0
k_1	Number of scheduling resources
k_2	Number of scheduling entities
L	Longest path in graph before the priority change
L'	Longest path in graph after the priority change
$L(n_j)$	Longest path that includes node n_j
$L_b(n_y, n_z)$	Longest path in G_b from node n_y to n_z
$L_f(n_x, n_y)$	Longest path in G_f from node n_x to n_y
m_j	Subpath in the graph associated to node n_j
M	Maximum number of possible priorities (durations) for passes with random priorities (durations)
M_c	Maximum number of passes that change priority at the same time
$\mathfrak{M}_i$	Machine (unified notation)
M_j	Number of possible priorities (durations) for pass p_j^w (request j_j^w)
n_{d_j}	Discrete due time of request j_j
n_{e_k}	Discrete end time of pass p_k
n_i	Node in the graph
n_{r_j}	Discrete release time of request j_j
n_{s_k}	Discrete start time of pass p_k
$n^*(i, k)$	Best node in Z_i for current stage Z_k with associated pass tracked
$n^0(i, k)$	Best node in Z_i for current stage Z_k with associated pass not tracked
N	Number of requests or passes
o_j	Visibility window j
$O(\cdot)$	Big O notation
$\mathfrak{p}_{ij}$	Processing time of job $\mathfrak{J}_j$ in machine $\mathfrak{M}_i$ (unified notation)
$\mathfrak{p}_{ij}^{var}$	Variable processing time for job $\mathfrak{J}_j$ in machine $\mathfrak{M}_i$ (unified notation)
$\overline{\mathfrak{p}}_j$	Maximum processing time of job $\mathfrak{J}_j$ (unified notation)
$\underline{\mathfrak{p}}_j$	Minimum processing time of job $\mathfrak{J}_j$ (unified notation)
$p_{j,m}$	Pass with associated failure probability $\alpha_{j,m}$ generated from p_j^w
p_j^w	Pass with random priorities
p_k	Pass k
$p(e_i)$	Pass associated to event e_i
$\mathfrak{P}$	Multiple related machines (unified notation)
P	Initial set of passes
P', P'', P_{sub}	Schedule
$\tilde{P}$	Executed schedule
P^*	Optimal schedule
$P^*(t)$	Updated optimal schedule
P_j^a	Set of passes that are active for node n_j
P^d	Set of passes generated from the set of requests with random durations
P^{dR}	Robust schedule for the robust SRS problem with random durations
P^f	Feasible schedule
P_j^j	Set of passes generated from request j_j

P_j^{w}	Set of passes with failure probabilities generated from p_j^{w}
P_l	Set of later passes for p_l, including p_l
P_l^{p}	Precedence subset
P_{nw}	Set of passes with uncertain priorities
P^{R}	Robust schedule for the robust SRS problem with failure probabilities
P_i^{s}	Subset of P in which passes are associated to s_i
P_i^{sm}	Security schedule for s_i
P^{w}	Initial set of passes with random priorities
P^{wR}	Robust schedule for the robust SRS problem with random priorities
$P(n_j)$	Set of passes associated to backtracking from n_j to n_0
$P(t)$	Set of passes with time-dependent priorities at time t
$P(t)\vert_{t_1}^{t_2}$	Subset of $P(t)$, which passes have start times in the interval $[t_1, t_2]$
$\mathbb{P}(\cdot)$	Probability
$\overline{\overline{q}}$	Maximum duration of all visibility windows in discretization
$\overline{q}_j$	Duration of visibility window of discretized request j_j
$\overline{q}_{jk}$	Maximum duration of pass p_k associated to discretized request j_j
$\underline{q}_{jk}$	Minimum duration of pass p_k associated to discretized request j_j
r_j	Release time of request j_j
$\mathfrak{r}_j$	Release time of job $\mathfrak{J}_j$ (unified notation)
$\mathfrak{R}$	Multiple unrelated machines (unified notation)
$R_i(k)$	Best pair of nodes in Z_i, computed from current stage Z_k
s_h	Satellite h
s_{lead}	Leader satellite
S	Set of satellites
$\S$	Section
t	Time
t_0	Initial time
t_{e_k}	End time of pass p_k
t_{s_k}	Start time of pass p_k
$\mathscr{T}$	Tracking pass action
T	Duration of the scheduling horizon
T_{g}	Reduced time for the game development
$\mathfrak{L}_j$	Lateness of job $\mathfrak{J}_j$ (unified notation)
$U[1, 10]$	Discrete uniform distribution with values between 1 and 10
v	Edge in (forward) graph
$\mathfrak{w}_j$	Priority of job $\mathfrak{J}_j$ (unified notation)
$w_{j,m}$	Priority of pass $p_{j,m}^{\mathrm{w}}$
w_k	Priority of pass p_k
$w_{\mathrm{th}}^k(p_z)$	Threshold value for the priority of pass p_z for current stage Z_k
w.p.	With probability
$W_i(k)$	Priorities associated to the best pair of nodes $R_i(k)$
W_j	Distribution of priorities for p_j^{w}

Z^+	Stages associated to start time events in graph
Z_0	Initial stage in graph
$\lVert \cdot \rVert_{\Sigma w}$	Metric of schedule, e.g., $\lVert P' \rVert_{\Sigma w}$
$\lVert \cdot \rVert_{\mathbb{E}}$	Expected metric of schedule, e.g., $\lVert P' \rVert_{\mathbb{E}}$
$\langle \cdot \rangle_{t_s}$	Set of passes sorted by start time, e.g., $\langle P' \rangle_{t_s}$
$\widehat{}$	Reduced information (unified notation), e.g., $\widehat{w}_j$, $\widehat{\mathfrak{U}}_j$
$\widetilde{}$	Uncertainty (unified notation), e.g., $\widetilde{p}_{ij}$, $\widetilde{\mathfrak{U}}_j$

List of Figures

Fig. 1.1   Book structure 6

Fig. 2.1   Scheduling scenario 12
Fig. 2.2   Scheduling process 12
Fig. 2.3   Visibility windows 13
Fig. 2.4   Scheduling requests (based on visibility windows from Fig. 2.3) .. 14
Fig. 2.5   Final schedule (based on requests from Fig. 2.4) 14
Fig. 2.6   Satellite Range Scheduling problems 16

Fig. 3.1   Time varying graph representing visibility between
pairs ground station–satellite 22
Fig. 3.2   Example of TVG modeling two visibility windows
for two ground stations and two satellites: interaction
between s_1 and g_2 only active at $\tau_{s_1} < t < \tau_{e_1}$ (*blue*),
and interaction between s_2 and g_1 only active at
$\tau_{s_2} < t < \tau_{e_2}$ (*green*) 23
Fig. 3.3   Request specification 24
Fig. 3.4   Types of requests: variable-slack (*top*), fixed-slack
(*center*), and no-slack (*bottom*) 26
Fig. 3.5   Transformation of a request into a set of passes, with all
the possible combinations of start times and durations
complying with the specification of the request, and
taking into account the duration of the discretization step 29
Fig. 3.6   Examples of conflicting passes 31
Fig. 3.7   Priorities of the passes generated from a request 33
Fig. 3.8   Unified notation for representing scheduling problems 38
Fig. 3.9   SRS problems classification 46

Fig. 4.1   Centralized scheduler 50
Fig. 4.2   Relations between the algorithms for optimal SRS 58
Fig. 4.3   Set of passes 61
Fig. 4.4   Event generation 62

Fig. 4.5   Transition from stage Z_0 to Z_1. (**a**) Initial node. (**b**)
            Creation of new node. (**c**) Frontier B_1 63
Fig. 4.6   Transition from stage Z_1 to Z_2. (**a**) Creation of new
            node. (**b**) Frontier B_2... 63
Fig. 4.7   Transition from stage Z_2 to Z_3. (**a**) Creation of new
            nodes. (**b**) Frontier B_3 64
Fig. 4.8   Transition from stage Z_3 to Z_4. (**a**) Creation of new
            node. (**b**) Frontier B_4... 65
Fig. 4.9   Graph generation example. [2] ©2014 Scitepress.................. 66
Fig. 4.10   Simulation scenarios. (**a**) Practical case. (**b**) Worst case........... 67
Fig. 4.11   Optimal schedule (*blue*) and dismissed passes (*orange*)
            for the practical case scenario for a 1 day scheduling horizon 68
Fig. 4.12   Simulation: practical case. (**a**) Simulation times. (**b**)
            Metric ratios. [2] ©2014 Scitepress................................. 68
Fig. 4.13   Optimal schedule (*blue*) and dismissed passes (*orange*)
            for the worst case scenario for a 1 day scheduling horizon 69
Fig. 4.14   Simulation: worst case. (**a**) Simulation times. (**b**)
            Metric ratios. [2] ©2014 Scitepress................................. 69
Fig. 4.15   Number of passes and scheduling horizon: extended
            practical case. (**a**) Varying the number of ground
            stations. (**b**) Varying the number of satellites...................... 70
Fig. 4.16   Number of passes and scheduling horizon: extended
            worst case. (**a**) Varying the number of ground stations.
            (**b**) Varying the number of satellites 71
Fig. 4.17   Availability of partial results. (**a**) Practical case. (**b**)
            Worst case.. 71
Fig. 4.18   Optimal solutions to the SRS problem, previous
            literature (*green*) and this book (*blue*) 72

Fig. 5.1   SRS scenarios with different topologies. (**a**) Centralized
            SRS. (**b**) Distributed SRS with perfect information.
            (**c**) Distributed SRS with uncertain priorities. (**d**)
            Distributed SRS with uncertain passes 78
Fig. 5.2   Distributed scheduler 79
Fig. 5.3   Relations between the algorithms.................................... 92
Fig. 5.4   Passes for the SRS game example 95
Fig. 5.5   Stage Z_1. (**a**) Graph generation. (**b**) Extensive form representation   96
Fig. 5.6   Stage Z_2. (**a**) Graph generation. (**b**) Extensive form representation   98
Fig. 5.7   Stage Z_3. (**a**) Graph generation. (**b**) Extensive form representation   98
Fig. 5.8   Stage Z_4. (**a**) Graph generation. (**b**) Extensive form representation   98
Fig. 5.9   Passes (*top*), graph (*middle*), and game in extensive
            form (*bottom*). [4] ©2015 IEEE.................................... 100
Fig. 5.10   Simulation results: scenario 1. (**a**) Simulation times.
            (**b**) Price of anarchy. [4] ©2015 IEEE 102

Fig. 5.11   Simulation results: scenario 2. (a) Simulation times.
            (b) Price of anarchy. [4] ©2015 IEEE   103
Fig. 5.12   Payoff differences. (a)–(e) Scenario 1 results. (f)–(j)
            Scenario 2 results. [4] ©2015 IEEE   103
Fig. 5.13   Noncooperative SRS and relations with centralized SRS...........   105

Fig. 6.1   Robust scheduler..   109
Fig. 6.2   First example: set of later non-conflicting passes for a
            single scheduling entity   111
Fig. 6.3   Second example: set of later non-conflicting passes for
            several scheduling entities...   111
Fig. 6.4   Passes associated to a multiply connected belief
            network. (a) Set of passes. (b) Associated belief network   115
Fig. 6.5   Passes generated from two identical requests which are
            associated to a multiply connected belief network. (a)
            Set of passes. (b) Associated belief network   122
Fig. 6.6   Relations between the algorithms for robust SRS...................   123
Fig. 6.7   Set of passes for a single ground station. [3] ©2015 IEEE   124
Fig. 6.8   Simulation results for scenario 1. (a) Expected metrics.
            (b) Metrics of executed schedules. [3] ©2015 IEEE................   125
Fig. 6.9   Simulation results for scenario 2. (a) Expected metrics.
            (b) Metrics of executed schedules. [3] ©2015 IEEE................   126
Fig. 6.10   Complexity of robust SRS and relations with
            deterministic SRS..   128

Fig. 7.1   Reactive scheduler   130
Fig. 7.2   Examples of scenarios for static and dynamic SRS:
            static SRS (top), dynamic SRS solved with static
            approach (center), and reactive SRS (bottom);
            indicating periods where computations are required for
            finding new/alternative schedules (blue), periods where
            the schedule is executing optimally with respect to the
            current set of priorities (green), and periods where the
            schedule is executing suboptimally after a change in
            priorities (red)...   131
Fig. 7.3   Preprocessing phases   133
Fig. 7.4   Set of passes   141
Fig. 7.5   Phase 1: forward graph generation...................................   141
Fig. 7.6   Phase 2: backward graph generation.................................   142
Fig. 7.7   Phase 3: best pairs in stages search   143
Fig. 7.8   Phase 4: alternative paths for pass 1   144
Fig. 7.9   Phase 4: alternative paths for pass 2   145
Fig. 7.10   Phase 4: alternative paths for pass 6   146

Fig. 7.11   Complexity of reactive SRS and relations with basic
            SRS and robust SRS 147
Fig. 8.1   Summary of SRS problems solved in this book (with
            Δt and FNE) 150

List of Tables

Table 3.1   Complexity of general and satellite scheduling. [9]
            (extended) ©2014 Springer 45

Table 5.1   Noncooperative SRS problems 105

Table 6.1   Optimal solutions for Robust SRS.................................. 127

Table 7.1   Optimal solutions for Reactive SRS 146

Part I
Introduction

Chapter 1
Motivation

The Satellite Range Scheduling (SRS) problem is an important problem in operations research for aerospace systems. In a nutshell, it can be described as the allocation of tasks among a set of orbiting objects (satellites) and a set of Earth-bound objects during visibility times. There are three main fields where SRS is applicable:

- *Satellite communications*, where tasks are communication intervals between sets of satellites and ground stations.
- *Earth observation*, where tasks are observations of spots on the Earth by satellites.
- *Sensor scheduling*, where tasks are observations of satellites by sensors on the Earth.

The progressive increase in the number of satellites and ground stations triggers a combinatorial explosion in the number of these intervals to be scheduled. For example, ESA's Tracking Station Network (ESTRACK) had 30 satellites and 40 antennas in 2010 [1], NASA's Deep Space Network (DSN) 35 satellites and 13 antennas (with 425 requests per week) in 2006 [2], and the Air Force Satellite Control Network (AFSCN) had 100 satellites and 16 antennas (with 500 requests per day) in 2006 [3]. Furthermore, some applications require taking into account uncertainty or dynamically introducing new tasks, which further complicates the problem.

1.1 Motivation

The scheduling of these tasks has been posed as an optimization problem since its inception, and has been studied for decades. Despite this, optimal solutions

© Springer International Publishing Switzerland 2015
A.J. Vázquez Álvarez, R.S. Erwin, *An Introduction to Optimal Satellite Range Scheduling*, Springer Optimization and Its Applications 106,
DOI 10.1007/978-3-319-25409-8_1

were known only for very special cases. This fact raises several questions: Is it possible to find the optimal schedule of the general problem in a computationally feasible way? What are the impacts if the satellites or ground stations do not follow the proposed solution? And what if the tasks fail or take longer than expected, could an optimal schedule be found under this uncertainty in a computationally feasible manner? And if the conditions of the problem change dynamically, can a dynamic scheduling approach be found to react to these changes?

Through the following chapters we will describe a model for characterizing the tasks to be scheduled, based on which we will formally present the scheduling problem. We will define the tractability bounds for the basic version of SRS, which is the most studied one, and will present also optimal solutions for some of its variants.

1.2   Why Optimal Scheduling?

Most of existing literature on SRS is focused on approximate solutions. Even if these algorithms provide acceptable levels of performance (e.g., total communication time), and relative comparison can be done among them (e.g., Algorithm 1 provides a 5 % increase in performance compared to Algorithm 2), the absence of a reference solution for doing absolute comparison prevents algorithm designers from knowing where the actual performance limits are.

Exhaustive search is dismissed for providing a reference solution, specially as it cannot be applied to large scale, complex scenarios. Having an optimal solution in tractable time (let us say for now polynomial under reasonable assumptions) provides a scheduling algorithm usable in real applications, or at a minimum a reference for benchmarking currently used heuristic/suboptimal approaches. It also provides a basis for looking at and solving variants of the problem, as will be shown in this book.

1.3   Why this Book?

Our intention with this work is to provide satellite operations engineers with a small handbook with reference algorithms for tackling SRS problems, classified according to the parameters most used in existing literature.

This book is the result of two years of research in the SRS problem. The lack of an optimal solution computable in tractable time, when tackling a distributed version of this problem, led us to the search of this optimal solution. The approach of this solution showed to be valid with some adjustments to some variants of this problem.

In this sense, this monograph is strongly based on our previous work, published during this research period [4–7], which has been considerably extended, including new variants of the problem, better structured, and integrated to provide the reader with the big picture of SRS, and further detailed to make the text more accessible to those who have never been exposed to SRS problems previously.

Of course this work is not a complete compendium of all the scheduling problems available in the literature, we are aware that we are only scratching the surface. However, we consider that our formal and systematic approach will also allow to solve more complex versions of the problems here presented. We also hope that this work saves time to SRS algorithm designers, no matter if taking their first steps in this field or designing algorithms for specific applications.

1.4 Structure of the Book

The book is divided into three parts and seven chapters:

Part I: Introduction. In this part we introduce this book, describe the scheduling process and provide an overview of the SRS problem and its main variants.

- In **Chap. 1: Motivation** we present the motivation for this monograph, and this description of the rest of the chapters.
- In **Chap. 2: Scheduling Process** we introduce the scheduling in context with the satellite mission, and present the parameters for the classification of SRS problems. This chapter is aimed at those readers not familiar yet with the scheduling process and its relation with planning in satellite missions.

Part II: Satellite Range Scheduling. In this part we present the mathematical formulation of the SRS problem, and provide its optimal solution.

- In **Chap. 3: The Satellite Range Scheduling Problem** we formally present the SRS problem. Reading this chapter is recommended before addressing any of the following chapters, as it provides the model for the intervals to be scheduled, and it presents several variations of this basic problem which can also be extended to those problems presented in the next chapters.
- In **Chap. 4: Optimal Satellite Range Scheduling** we present a dynamic programming algorithm for solving the fixed interval SRS problem, and compare its performance to traditional algorithms. We also present an extension of this algorithm for solving the general case, with variable durations for the requests and discretized time.

Part III: Variants of Satellite Range Scheduling. In this part we provide three variants of the SRS problem: distributed, stochastic, and dynamic.

- In **Chap. 5: Noncooperative Satellite Range Scheduling** we present a distributed formulation of the problem, where rather than providing a centralized schedule for the system, scheduling is performed at either the satellites or

the ground stations in a distributed fashion. We model the problem through
a noncooperative game-theoretic perspective, so that we will assume that the
entities aim at optimizing their own schedules selfishly. We consider different
versions of the problem regarding the information available to each operator.

- In **Chap. 6: Robust Satellite Range Scheduling** we introduce uncertainty in
  the model of the problem. We consider that the communication intervals are
  subject to failure with a certain probability, so that the objective will be to
  provide a robust schedule which maximizes the expected performance. We use
  this framework to model more complex versions of the problem which take into
  account uncertainty into the priorities or the durations of the communications.
- In **Chap. 7: Reactive Satellite Range Scheduling** we present a different formulation of the problem in which the priorities of the communication intervals are
  known at all times but subject to change. Rather than providing a static schedule
  that maximizes the expected performance, we provide an algorithm for speeding
  up the computation of the optimal schedule.
- In **Chap. 8: Summary** we present a summary showing the relations between the
  problems that have been solved in this book. This last chapter provides the reader
  with the big picture for the presented SRS problems, and will allow to understand
  where efforts should be devoted in future work.

Even though Chap. 2 may be skipped by those familiar with scheduling in
satellite missions, this chapter will provide more detail on the structure of the
book, and therefore its reading is encouraged. Figure 1.1 displays the dependencies
between chapters.

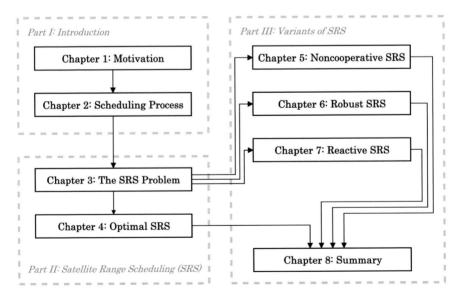

Fig. 1.1 Book structure

1.5 Main Contributions

We present the main contributions for each chapter, highlighting those contributions which are new to this book compared to the referenced articles:

Chapter 1 *New chapter.*
Included introduction and structure of this book.

Chapter 2 *New chapter.*
Included description of the scheduling process and introduction to the problems tackled in this book.

Chapter 3 *Strongly based on [4].*
Extended explanations and included examples supporting the definitions for the model of the scenario (§3.1.1), the model of the requests (§3.1.2) and the problem constraints (§3.1.3).
Provided a more precise definition of redundancy, more related with satellites or ground stations with multiple capacity (§3.1.3).
Improved the definition and explanations for precedence (§3.1.3).
Included introduction to complexity theory (§3.2.1).
Simplified proofs and included new results for precedence (§3.2.2).
Extended explanations for the problem classification, and included new cases with precedence and redundancy constraints (§3.3).
Extended summary with more problems and references, and added representation of the relations of the presented problems (§3.5).

Chapter 4 *Strongly based on [5].*
Simplified the model for the scenario and extended explanations based on previous chapter (§4.1).
Extended the explanations on the development of the graph (§4.2.1).
Included extension of the algorithm for solving the slack (variable duration) version of the problem with discretized time (§4.3.1).
Included extension of the algorithm for solving the variant of the problem with redundancy (§4.3.2).
Included remarks on the complexity of the presented algorithm compared to traditional SRS algorithms (§4.4).
Detailed generation of the graph for the presented example (§4.5).
Extended explanations for the simulated models, including representations of the scenario and the schedule (§4.6).
Included results on the relation between the number of passes and the length of the scheduling horizon (§4.6.3).
Included results on the availability of partial results for the presented algorithm compared to traditional algorithms (§4.6.4).
Included graphical representation of the available optimal solutions for SRS problems (§4.7).

Chapter 5 *Strongly based on [6].*
Extended explanations for the model of the scenario (§5.1).

Included variant of the problem with uncertain priorities, and added explanations relating the two game problems with the centralized topology version of the problem (§5.4.1).

Included variant of the problem with uncertain passes (§5.4.2).

Included explanations of some concepts associated to game theory which are applicable to the problems studied in this chapter (§5.5).

Detailed generation of the graph for the presented example (§5.6).

Extended notation introduced in previous chapters to cover the distributed SRS problems (§5.8).

Included a summary of all the presented problems, and a graphical representation of the relations between these problems (§5.8).

Chapter 6 *Strongly based on [7].*

Simplified scenario model and linked it with previous chapters (§6.1).

Added explanations and examples for the defined sets of passes (§6.2).

Included characterization for the multiple entity problem (§6.3).

Included variant of the problem with random priorities (§6.4.1).

Included variant of the problem with random durations (§6.4.2).

Included an extension of the provided algorithm to provide a faster solution for the single entity deterministic problem (§6.5).

Extended notation introduced in previous chapters to cover the robust SRS problems (§6.8).

Included a summary of all the presented problems, and a graphical representation of the relations between these problems (§6.8).

Chapter 7 *New chapter.*

Characterized the complexity of the reactive SRS problem (§7.1.1).

Provided an algorithm for the optimal solution of a restricted case minimizing the computation time through preprocessing (§7.2).

Provided an algorithm for reducing the recomputation in a generalization of the previous case (§7.3).

Provided an example on the preprocessing and recomputation for the restricted case (§7.4).

Included a summary of all the presented problems, and a graphical representation of the relations between these problems (§7.5).

Chapter 8 *New chapter.*

Included a summary of all the problems solved in this book (§8.1).

Included future lines of this work (§8.2).

Acknowledgements This research was performed while the author held a National Research Council Research Associateship Award at the Air Force Research Laboratory (AFRL).

References

1. W. Heinen, M. Unal, Scheduling tool for ESTRACK ground station management, in *SpaceOps 2010* (American Institute of Aeronautics and Astronautics, Reston, VA, 2010)
2. M.D. Johnston, B.J. Clement, Automating Deep Space Network scheduling and conflict resolution, in *Autonomous Agent and Multiagent Systems, AAMAS'06*, ACM (2006)
3. L. Barbulescu, A.E. Howe, L.D. Whitley, M. Roberts, Understanding algorithm performance on an oversubscribed scheduling application. J. Artif. Intell. Res. **27**, 577–615 (2006). AI Access Foundation
4. A.J. Vazquez, R.S. Erwin, On the tractability of satellite range scheduling. Optim. Lett. **9**(2), 311–327 (2015)
5. A.J. Vazquez, R.S. Erwin, Optimal fixed interval satellite range scheduling, in *Proceedings of the 3rd International Conference on Operations Research and Enterprise Systems* (Scitepress, Angers, 2014), pp. 401–408
6. A.J. Vazquez, R.S. Erwin, Noncooperative satellite range scheduling with perfect information, in *2015 IEEE Aerospace Conference* (IEEE, Big Sky, 2015)
7. A.J. Vazquez, R.S. Erwin, Robust fixed interval satellite range scheduling, in *2015 IEEE Aerospace Conference* (IEEE, Big Sky, 2015).

Chapter 2
Scheduling Process

In this chapter we put the scheduling process in the context of the various missions that the problem arises in. This chapter is intended for readers that are approaching SRS, or for professionals from other industries, to show where all the constraints of this problem come from. As we mentioned in the previous chapter, SRS is applicable in satellite communications, Earth observation, and sensor scheduling. We will describe the scheduling process applied to satellite communication scenarios, without loss of generality for other cases.

Although the experienced professional may skip this chapter, reading of §2.2 and §2.3 is encouraged as these subsections present the parameters used in this book to classify scheduling problems.

2.1 Scheduling Process

For our scheduling model we consider a set of *ground stations* which move with the surface of the Earth, a set of *mission control centers* which can be assumed to be continuously connected to the ground stations, and *satellites* traveling through different kinds of orbits generating *visibility windows* when line of sight (LOS) to ground stations exist [1]. A possible scenario is displayed in Fig. 2.1.

The satellite *operators* aim to establish communications between their mission control center and their associated satellite, but this can only be done through the *ground station network*. Based on the dynamics of the scenario and on the requirements of the mission, the operators generate an operations plan. Then, from these visibility time windows and the operations plan, the operators generate a set of *requests* characterized by constraints associated to these time windows. The

© Springer International Publishing Switzerland 2015
A.J. Vázquez Álvarez, R.S. Erwin, *An Introduction to Optimal Satellite Range Scheduling*, Springer Optimization and Its Applications 106,
DOI 10.1007/978-3-319-25409-8_2

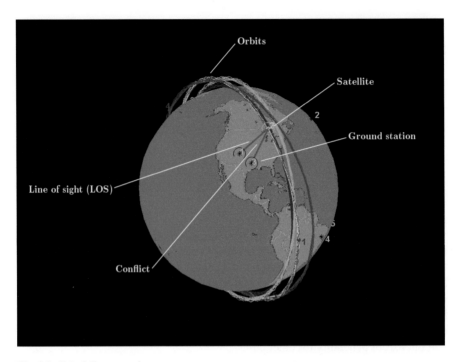

Fig. 2.1 Scheduling scenario

Fig. 2.2 Scheduling process

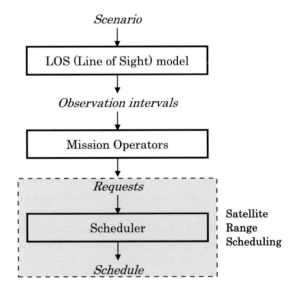

objective of the scheduler is to generate, from this set of requests, a *schedule*, which is a subset of these requests selected for execution. This process is illustrated in Fig. 2.2.

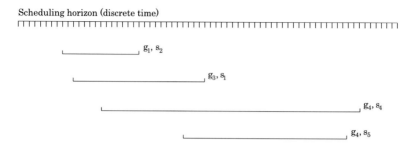

Fig. 2.3 Visibility windows

   The duration of the visibility windows will depend on the dynamics of the
problem (orbits of the satellites and positions of the ground stations). For a low
Earth orbit (LEO) satellite these durations may be of around 10 min at most. Given
that this time is relatively small, the whole visibility window is generally requested.
For higher orbits however, these durations may be longer than one hour, up to the
case of geostationary orbits (GEO), where there may be continuous visibility. For
this last case the operators may only need a small section of the visibility window.
This section will be delimited by a *release time* and a *due time* (terms widely used
in general machine scheduling referring to the availability and deadline times of the
tasks), inside which the operators will request a *minimum duration* and a *maximum
duration* of communication with the satellite.
   We will consider problems that require scheduling over a finite time duration,
termed the *scheduling horizon*, that is, finite time limits for enclosing the set of
requests to be scheduled, e.g., seven days [2, 3]. We furthermore consider time to
be discretized with a certain *discretization step*, e.g., 1 min [2, 3]. We show a set of
visibility windows in Fig. 2.3.
   The constraints on the requests are not only time-related. The operators may
prefer to communicate with the satellite under lighting conditions, or more critically,
the operators may need to communicate as soon as possible. Therefore these
requests will have associated *priorities* to model these preferences. Reference [4]
introduces the use of a suitability function for calculating these priorities.
   Additionally, the operations plan may require one of the communications to be
performed before another one. This constraint is known as *precedence* (imagine for
example a relay network). Another constraint is that the communication may be
required to be performed without interruptions, that is, with no *preemption*. And in
some cases, two or more ground stations may communicate at the same time with
the same satellite, which is known as *redundancy* or multiple (*m*-ary) capacity. Otherwise the problem has unitary capacity, and time-overlaying requests associated to
either the same satellite or ground station are considered a *conflict*. For this case with
unitary capacity a schedule with conflicts will not be *feasible*, and thus neither it
will be valid as a possible solution to the problem. Generally no-preemption and no-
redundancy are considered [2, 3], although some references do consider redundancy
[5]. We will provide formal definitions for these terms in Chap. 3 (§3.1.2 and §3.1.3),
but they are briefly presented now to illustrate the problem.

The Satellite Range Scheduling (SRS) problem requires us to find a feasible schedule that maximizes the sum of the priorities of the requests included in this schedule, given the requests and the associated constraints. We will refer to this sum of priorities as the performance or *metric* of the schedule. The problem will be even more complicated as we have several satellite missions to be served in a ground station network [6–8].

Additional constraints are introduced in the previous description which define different variants of this problem. In some systems, the mission operators may schedule in a *distributed* fashion rather than following a centralized schedule (consider for example different missions with different objectives computing their associated schedules without a central coordinating entity). In other cases, there could be *uncertainty* in the duration of the communication (e.g., considering uncertainty on the total amount of data to be sent). And other scenarios may be *dynamic*, requiring to be able to quickly react to changes in the model of the requests (consider the case where a satellite enters safe mode and requires immediate communication). These three variants are explained in more detail in Chaps. 5, 6 and 7, respectively.

We show in Fig. 2.4 a set of requests, indicating the start and end times of the visibility windows, the minimum and maximum duration requested by the operators along with their associated priorities, and the relations of no-preemption, no-redundancy, and precedence. In Fig. 2.5 we show a feasible schedule, complying with the constraints specified in Fig. 2.4.

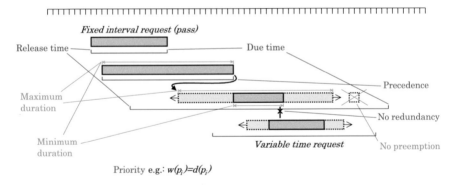

Fig. 2.4 Scheduling requests (based on visibility windows from Fig. 2.3)

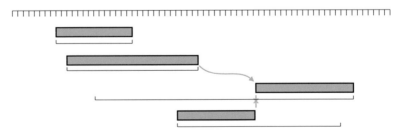

Fig. 2.5 Final schedule (based on requests from Fig. 2.4)

2.2 Scheduler Characteristics

We describe the different kinds of schedulers that can be used to solve the problem, and which depend on additional constraints on the missions. Our classification involves three parameters generally used in existing literature:

- *Topology*: the calculation of the schedule may be performed in a *centralized* fashion (that is, a single schedule is calculated based on the information from all the entities), or in a *distributed* fashion (different entities compute their associated schedules independently, without a coordinating entity).
- *Uncertainty*: we say that the problem is *deterministic* if the entities are 100% reliable, or that it is *stochastic* if the reliability is less than 100% (e.g., scheduled requests have a probability of not being executed).
- *Changes*: we say the requests are *static* if they do not change from the start of the scheduling horizon to its end, and that requests are *dynamic* if they may change before the completion of the scheduling horizon.

2.3 Satellite Range Scheduling Problems

According to the classification presented in §2.2, we present the problems that we will tackle in this book:

- *Satellite Range Scheduling*: classified as centralized, deterministic, and static, in this case the objective is to find the optimal schedule. We present the formulation of the problem in Chap. 3, and provide an algorithm for finding the optimal schedule in Chap. 4.
- *Noncooperative Satellite Range Scheduling*: classified as distributed, deterministic, and static, in this case the scheduling entities (either the ground stations or the satellites) aim at maximizing their own performance. We tackle this problem through a game-theoretic approach in Chap. 5.
- *Robust Satellite Range Scheduling*: classified as centralized, stochastic, and static, in this case the communication intervals are subject to fail, so that an approach to provide the maximum expected performance will be presented in Chap. 6.
- *Reactive Satellite Range Scheduling*: classified as centralized, deterministic, and dynamic, in this case requests change along the execution of the window, so the computation of the optimal schedule needs to be speed up. An approach based on the solution of the basic SRS problem is presented for this problem in Chap. 7.

A graphical representation for this classification is shown in Fig. 2.6.

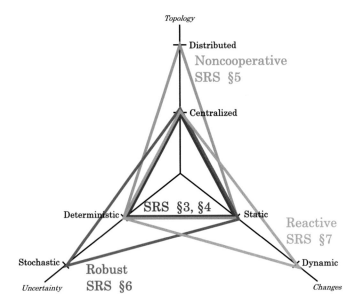

Fig. 2.6 Satellite Range Scheduling problems

2.4   Issues Beyond the Scope of this Text

As we have stated previously, this book is focused on reference (optimal) solutions. Therefore it is out of the scope of this work to reference lists of available *suboptimal solution algorithms* (see, for example, [2, 9, 10] for enumeration and performance comparison).

Neither we consider *satellite-specific models* including power, memory, and channel capacity (for suboptimal solutions on these problems see for example [10] for the deterministic problem with a single resource, [11] for the stochastic problem with failure probabilities, and [12] for the dynamic problem).

It is also out of the scope of this book to detail the *scheduling process in professional networks* (see [7, 13–15] for scheduling in NASA's Deep Space Network (DSN), [6, 16–18] for the ESA Tracking Station Network (ESTRACK), and [8, 19] for the Air Force Satellite Control Network (AFSCN)).

Acknowledgements This research was performed while the author held a National Research Council Research Associateship Award at the Air Force Research Laboratory (AFRL).

References

1. D.A. Vallado, *Fundamentals of Astrodynamics and Applications* (Space Technology Library, Microcosm Press, Portland, OR, 2001)
2. L. Barbulescu, J.P. Watson, L.D. Whitley, A.E. Howe, Scheduling space-ground communications for the air force satellite control network. J. Sched. **7**(1), 7–34 (2004)
3. F. Marinelli, F. Rossi, S. Nocella, S. Smriglio, A Lagrangian heuristic for satellite range scheduling with resource constraints. Comput. Oper. Res. **38**(11), 1572–1583 (2005)
4. W.J. Wolfe, S.E. Sorensen, Three scheduling algorithms applied to the Earth observing systems domain. Manag. Sci. **46**(1), 148–168 (2000)
5. M. Schmidt, Ground station networks for efficient operation of distributed small satellite systems. Ph.D. Thesis, University of Wurzburg, 2011
6. W. Heinen, M. Unal, Scheduling tool for ESTRACK ground station management, in *SpaceOps 2010* (American Institute of Aeronautics and Astronautics, Reston, VA, 2010)
7. M.D. Johnston, B.J. Clement, Automating Deep Space Network scheduling and conflict resolution. *Autonomous Agent and Multiagent Systems, AAMAS'06*, ACM (2006)
8. L. Barbulescu, A.E. Howe, L.D. Whitley, M. Roberts, Understanding algorithm performance on an oversubscribed scheduling application. J. Artif. Intell. Res. **27**, 577–615 (2006)
9. A. Globus, J. Crawford, J. Lohn, A. Pryor, A comparison of techniques for scheduling Earth observing satellites, in *Proceedings of the Sixteenth Innovative Applications of Artificial Intelligence Conference. IAAI* (2004)
10. S. Spangelo, Modeling and optimizing space networks for improved communication capacity. Ph.D. Thesis. University of Michigan, 2013
11. J. Wang, E. Demeulemeester, D. Qiu, A pure proactive scheduling algorithm for multiple Earth observation satellites under uncertainty of clouds. SSRN (2014). doi:10.2139/ssrn.2495339
12. J. Wang, X. Zhu, D. Qiu, L.T. Yang, Dynamic scheduling for emergency tasks on distributed imaging satellites with task merging. IEEE Trans. Parallel Distrib. Syst. **25**(9), 2275–2285 (2014)
13. B.J. Clement, M.D. Johnston, The Deep Space Network scheduling problem, in *IAAI'05 Proceedings of the 17th Conference in Innovative Applications of Artificial Intelligence*, **3** (AAAI, Palo Alto, CA, 2005), pp. 1514–1520
14. M.D. Johnston, D. Tran, B. Arroyo, S. Sorensen, P. Tay, B. Carruth, A. Coffman, M. Wallace, Automating mid- and long-range scheduling for NASA's Deep Space Network, in *SpaceOps 2012* (American Institute of Aeronautics and Astronautics, Reston, VA, 2012)
15. M.D. Johnston, D. Tran, Automated scheduling for NASA's Deep Space Network, in *6th International Workshop on Planning and Scheduling in Space* (Space Telescope Science Institute, Baltimore, MD, 2011)
16. S. Damiani, H. Dreihahn, J. Noll, M. Niezette, G.P. Calzolari, Automated allocation of ESA ground station network services, in *International Workshop on Planning and Scheduling for Space* (American Association for Artificial Intelligence, Palo Alto, CA, 2006)
17. S. Damiani, H. Dreihahn, J. Noll, M. Niezette, G.P. Calzolari, A planning and scheduling system to allocate ESA ground station network services, in *The International Conference on Automated Planning and Scheduling, ICAPS 2007* (AAAI, Palo Alto, CA, 2007)
18. H. Dreihahn, M. Niezette, M. Gotzelmann, Centralized schedule and SLE service configuration file generation with the ESTRACK scheduling system. *European Ground System Architecture Workshop* (ESA, Paris, 2007)
19. B.R. Hays, A.M. Carlile, T.S. Mitchel, Visualizing and integrating AFSCN utilization into a common operational picture. *Advanced Maui Optical and Space Surveillance Technologies Conference, AMOS 2006* (Maui Economic Development Board, Kihei, HI, 2006)

Part II
Satellite Range Scheduling

Chapter 3
The Satellite Range Scheduling Problem

As introduced in Chap. 1, scheduling of the interactions between spacecraft and Earth-bound entities arises in multiple applications, including remote sensing (highly related to Earth observation satellites (EOS) [1]) and satellite communications (mainly those based on ground station networks (GSN) [2]).

Several instances of this general problem have been studied for the last decades [1–7], but different notations and vocabulary make comparing techniques difficult. This has been noticed for general scheduling [8], and existing bibliography provided some steps on this generalization of the notation [1–4].

In this chapter we establish a precise mathematical definition of the SRS problem, identify the SRS problem's computational complexity, and provide bounds on the computational scaling of the problem. We also complete the unification of the notations from satellite and general scheduling, started for simplified versions of the problem in previous work, and survey existing and new relations among problems from both fields.[1]

3.1 Problem Formulation

In this section we present a mathematical formulation of the SRS problem for supporting the definitions of the most important variants.

---

[1]This chapter is strongly based on our work [9], which has been extended, reorganized, and integrated with the rest of the book. For a detailed list of the new contributions please see §1.5.

© Springer International Publishing Switzerland 2015
A.J. Vázquez Álvarez, R.S. Erwin, *An Introduction to Optimal Satellite Range Scheduling*, Springer Optimization and Its Applications 106,
DOI 10.1007/978-3-319-25409-8_3

3.1.1 Model for the Scenario

Let t_0 be a time instant and T a time window (*scheduling horizon*) such that $t \in [t_0, t_0 + T]$. Let $S = \{s_h\}$ be a set of satellites, and $G = \{g_i\}$ a set of ground stations. For simplicity, we will only refer to these two kinds of entities (satellites and ground stations) through the rest of the book. As we noted in the previous chapter, the results hold for EOS scheduling problems, where the entities would be "observation satellites" and "ground targets," as well as sensor scheduling problems, where the entities would be "space objects" and "ground sensors," respectively.

Whereas ground stations can be considered to be moving with the surface of the Earth, satellites travel through different kinds of orbits. These two different motion dynamics generate visibility time windows when lines of sight (LOS) between satellites and ground stations exist [10].

Based on these visibility windows, the system can be modeled as a time varying graph (TVG), which consists of sets of nodes connected by edges whose presence varies along time. Figure 3.1 represents the visibility among ground stations and satellites as a time varying graph, with the links among pairs satellite–ground station being active according to the times when there is line of sight between the elements of these pairs. Except for some specific cases, the period of the Earth's rotation and the period of the satellite orbit, combined with perturbation effects which cause dynamic drift in the geometry of the orbit with respect to the Earth, prevent generating an analytic expression for when the LOS will exist; in the general case, the equations of motion are numerically integrated for a window into the future over which model errors and numerical errors can be kept bounded, and the results of these are used to compute when LOS will exist. This process justifies our use of a finite scheduling horizon later in the book.

The interactions (observation or communication) are limited to occur within these visibility windows, and could also involve time constraints specified by the

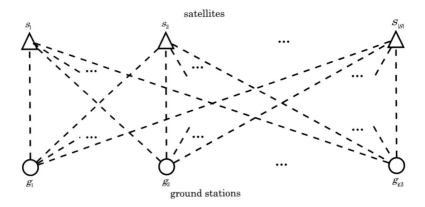

Fig. 3.1 Time varying graph representing visibility between pairs ground station–satellite

operators of the satellites, which depending on the specific mission may require fixed or variable contact times lasting for the entire window duration or just a portion of it. We illustrate the aforementioned model with an example in a system with two satellites and two ground stations, and with two visibility windows o_1 (involving s_1 and g_2) and o_2 (involving s_2 and g_1), limited by a pair of start and end times $o_j = (\tau_{s_j}, \tau_{e_j}) \; \forall j$. Let us suppose that $\tau_{s_1} < \tau_{s_2} < \tau_{e_1} < \tau_{e_2}$. Then we could represent the system as in Fig. 3.2.

We present in this chapter the baseline for the satellite range scheduling problem, and thus we start with the basic problem, where the topology is *centralized* so that the schedule can be computed once and for all the entities, which will follow it; and the interactions, characterized as requests, remain *deterministic* and *unchanged* along the scheduling horizon.

3.1.2   Model for the Requests

The operators of the mission will define a set of requests for the interactions among the satellites and the ground stations, which will be constrained to occur inside the visibility windows described in the previous section. Let r_j and d_j be the release and due times of a request associated to a visibility window $o_j = (\tau_{s_j}, \tau_{e_j})$, then:

$$\tau_{s_j} \leq r_j < d_j \leq \tau_{s_j}. \tag{3.1.1}$$

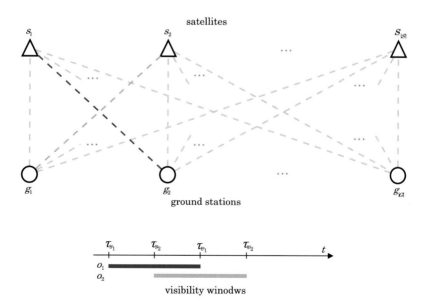

Fig. 3.2 Example of TVG modeling two visibility windows for two ground stations and two satellites: interaction between s_1 and g_2 only active at $\tau_{s_1} < t < \tau_{e_1}$ (*blue*), and interaction between s_2 and g_1 only active at $\tau_{s_2} < t < \tau_{e_2}$ (*green*)

The operators will define the minimum time they need over each request ρ_j and the maximum time $\overline{\rho_j}$ inside the interval $[r_j, d_j]$, and they will associate a priority to the interaction depending on the position inside the visibility window and its duration. We define formally the request as follows:

Definition 1. Let a *request* j_j be a tuple modeling an interaction request inside the visibility window defined by a release and a due time (r_j and d_j), and constrained by a minimum and a maximum duration (ρ_j and $\overline{\rho_j}$) between the satellite s_h and the ground station g_i, that is:

$$j_j \triangleq (s_h, g_i, r_j, d_j, \rho_j, \overline{\rho_j}, f_j^w(t)),\qquad (3.1.2)$$

where the indexes $h, i, j, N \in \mathbb{N}$, with $h \in [1, |S|], i \in [1, |G|], j \in [1, N]$, the lower and upper bounds for the durations $\rho_j, \overline{\rho_j} \in \mathbb{R}$ subject to $0 < \rho_j \leqslant \overline{\rho_j} \leqslant d_j - r_j$. The term $f_j^w(t)$ characterizes the (possibly time-varying) weight or priority (we use these two terms interchangeably) associated to that request, normalized between 0 and 1.

Figure 3.3 shows the release and due times, minimum and maximum duration, and priority of a request (function introduced in [1]), relating them with specific parameters like the maximum and minimum available data to be transmitted, the quality of the link, and the LOS acquisition and loss times. Throughout the rest of the book we will abstract from these specific parameters, as already noted in §2.4.

Let $J = \{j_j\}$ be a set of N interaction (communication or sensing) requests. This set of requests will be the input of the SRS problem.

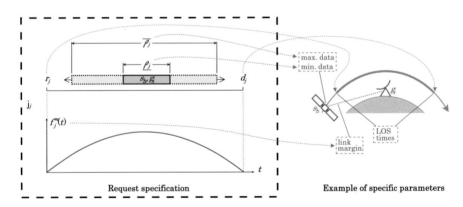

Fig. 3.3 Request specification

3.1.3 Problem Constraints

In this subsection we describe the constraints most known in the scheduling literature:

Preemption   Whether requests can be interrupted or not.
Number of entities   Whether the ratio between satellites and ground stations is one-to-many or many-to-many.
Duration of the requests   Whether the requested time is fixed or variable inside the visibility window.
Redundancy   Whether we allow one-to-one or one-to-many interactions.
Precedence   Whether some requests need others completed before being initiated or not.
Priority   Whether requests have the same or different priorities.

In the following subsections we define these parameters and show the relations between the possible cases for each of them.

3.1.3.1 Preemption

The allowance for preemption in scheduling problems means that requests can be interrupted while being served to attend a different request. This case is generally dismissed in the scheduling of satellites (see, for example, [1, 3]).

We will not delve here into the justification for this assumption, but it is generally related to long times required by the ground stations before starting to track the pass, basically because of the setup times required to orient the antennas and prepare the equipment (see, for example, [11]). The simplest approach is just taking into account these setup times inside the minimum durations of the requests. Therefore, the remainder of this text will make the assumption that no-preemption is allowed, and therefore requests cannot be interrupted once their servicing has started.

3.1.3.2 Number of Entities

A general classification is done in the literature depending on the number of scheduling entities ($|G|$ and $|S|$).

Definition 2. *Single resource range scheduling problems (SiRRSP) apply to those scenarios where there is only one satellite or only one ground station, multiple resource range scheduling problems (MuRRSP) to those where there are both multiple satellites and ground stations:*

$$\text{SiRRSP} \Leftrightarrow (|S| = 1) \vee (|G| = 1), \qquad (3.1.3)$$

$$\text{MuRRSP} \Leftrightarrow (|S| \geq 1) \wedge (|G| \geq 1). \qquad (3.1.4)$$

From this definition, the SiRRSP can be seen as a subproblem of the MuRRSP.

3.1.3.3   Duration of the Requests

Existing literature provides a general classification regarding the values of the required durations for the requests ($\underline{p_j}$ and $\overline{p_j}$), defined as minimum and maximum duration.

Definition 3. In the *no-slack* case the requested duration of the passes extends through the whole visibility window. In the *fixed-slack* case, the requested duration of the passes is fixed for each pass, but it can be smaller than the visibility window. And in the *variable-slack* case, the requested duration of the passes varies between two specified boundaries for each pass. According to the current notation:

$$\text{No-slack} \Leftrightarrow 0 < \underline{p_j} = \overline{p_j} = d_j - r_j \ \forall j_j \in J, \tag{3.1.5}$$

$$\text{Fixed-slack} \Leftrightarrow 0 < \underline{p_j} = \overline{p_j} \leqslant d_j - r_j \ \forall j_j \in J, \tag{3.1.6}$$

$$\text{Variable-slack} \Leftrightarrow 0 < \underline{p_j} \leqslant \overline{p_j} \leqslant d_j - r_j \ \forall j_j \in J. \tag{3.1.7}$$

From this definition the no-slack case is a subproblem of the fixed-slack case, which is a subproblem of the variable-slack case.

Figure 3.4 shows an example with the three types of requests, where j_1 is a variable-slack request (3.1.7), j_2 is a fixed-slack request (3.1.6), and j_3 is a no-slack request (3.1.5).

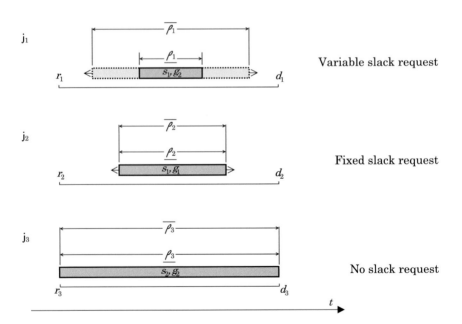

Fig. 3.4 Types of requests: variable-slack (*top*), fixed-slack (*center*), and no-slack (*bottom*)

Some references [7] associate the fixed-slack case to requests on high altitude orbits, and the no-slack one (also known as *fixed-interval scheduling* [8]) to requests on low altitude orbits. For high orbits (e.g., GEO) there may be visibility windows extending for the complete scheduling horizon, motivating the use of slack requests, whereas for low orbits (e.g., LEO) the short duration of the visibility windows generally makes necessary to extend the interactions through the whole visibility windows. These fixed interval requests are generally known as passes, which we define as follows:

Definition 4. Let a *pass* p_k be a tuple modeling a visibility time window from a start time t_s to an end time t_e between the satellite s_h and the ground station g_i, and with an assigned priority w_k, that is

$$p_k = (s_h, g_i, t_{s_k}, t_{e_k}, w_k) \ : \ w_k \in \mathbb{R} \cap [0, 1]. \tag{3.1.8}$$

The term w_k characterizes the weight or *priority* associated to this pass, normalized between 0 and 1.

We present a transformation which generates a set of passes from the initial set of requests (which is a generalization of the one presented in [1]), to build a common framework for the three slack cases. Let Δt be the discretization step, and Δt^{-1} the inverse of Δt.

Transformation 1. Every request j_j will be associated with a set $P_j^j = \{p_k\}_j$ of *passes*. Let D_n describe this association:

$$D_n : j_j \longrightarrow P_j^j = \{p_k\}_j \ : \ p_k \triangleq (s_h, g_i, n_{s_k}, n_{e_k}, w_k), \tag{3.1.9}$$

where s_h and g_i are the same as in the originating request j_j; $n_{s_k}, n_{e_k}, q_{j_k} \in \mathbb{Z}^*$ satisfy $n_{e_k} = n_{s_k} + q_{j_k}$ with $\lfloor \Delta t^{-1} r_j \rfloor \leqslant n_{s_k} \leqslant n_{e_k} \leqslant \lfloor \Delta t^{-1} d_j \rfloor$; $w_k \in \mathbb{R} \cap [0, 1]$.

Then, the pass p_k is a tuple entailing a satellite s_h and a ground station g_i (both taken from j_j), discrete start and end times (which are the integers n_{s_k} and n_{e_k}), a duration q_{j_k} (which is the difference among these discrete times), and a weight w_k. Thus the set P_j^j is the set of all the passes whose discrete start and end times comply with the conditions on the release and due times and with the durations constraints established in the request j_j following (3.1.2).

Proposition 3.1. *The transformation of the space of discrete slack requests into discrete passes is polynomial in the number of requests.*

Proof. For practical purposes let us assume, without loss of generality, that the values w_k are obtained in polynomial time from a discretized version of the suitability function $f_j^w(n\Delta t) \ \ \forall n \in \mathbb{Z}^*$. If this were not true, we could modify $f_j^w(n\Delta t)$ to fit this constraint since no optimality claims have been stated.

Let $n_{r_j} = \lfloor \Delta t^{-1} r_j \rfloor$ and $n_{d_j} = \lfloor \Delta t^{-1} d_j \rfloor$ be the release and due discrete times for a request j_j, and q_{j_k} the duration of the pass $p_k \in P_j^j$. The number of windows p_k associated to each request j_j can be obtained [1], from Transformation 1:

$$|P_j^i| = \sum_k (n_{d_j} - n_{r_j} - (q_{j_k} - 1)). \qquad (3.1.10)$$

Let $\overline{\overline{q}}_j = n_{d_j} - n_{r_j}$ and $\overline{\overline{\rho}}_j = d_j - r_j$. The result of the summation (3.1.10) can be easily solved [1] (n_{d_j} and n_{r_j} are constant values, so the sum reduces to an arithmetic finite series of q_{j_k}), where the low and high bounds for the sizes of the windows q_{j_k} for the worst case (variable-slack) are $\underline{q}_{j_k} = 1$ and $\overline{q}_{j_k} = \overline{\overline{q}}_j \ \forall k$ is:

$$|P_j^i| = \tfrac{1}{2} \left(\lfloor \Delta t^{-1} \overline{\overline{\rho}}_j \rfloor^2 + \lfloor \Delta t^{-1} \overline{\overline{\rho}}_j \rfloor \right). \qquad (3.1.11)$$

Applying Transformation 1 to the set J will generate a group $P = \{P_j^i\}$. Let $\overline{\overline{q}} = \max(\overline{\overline{q}}_j)$ and $\overline{\overline{\rho}} = \max(\overline{\overline{\rho}}_j)$. Then:

$$|P| = \sum_j |P_j^i|, \qquad (3.1.12)$$

$$|P| \leq \tfrac{1}{2}N \left(\lfloor \Delta t^{-1} \overline{\overline{\rho}} \rfloor^2 + \lfloor \Delta t^{-1} \overline{\overline{\rho}} \rfloor \right). \qquad (3.1.13)$$

Applying the slack definitions (Definition 3) on the sum (3.1.10) allows us to calculate the order of the transformation for every case as in inequality (3.1.13). Let J_{VS}, J_{FS}, and J_{NS} be the sets of requests constrained to variable-slack, fixed-slack, and no-slack, respectively. Then:

$$|D_n(J_{VS})| = O\left(N \lfloor \Delta t^{-1} \overline{\overline{\rho}} \rfloor^2\right), \qquad (3.1.14)$$

$$|D_n(J_{FS})| = O\left(N \lfloor \Delta t^{-1} \overline{\overline{\rho}} \rfloor\right), \qquad (3.1.15)$$

$$|D_n(J_{NS})| = O(N). \qquad (3.1.16)$$

And thus, the transformation D_n is polynomial in the number of requests N. □

Let $P = \{P_j^i\}$ be the space of all the possible *passes* (or interaction requests with fixed times), and every subset $P_{\text{sub}} \subseteq P$ a *schedule* (or set of passes to be tracked by the network). Figure 3.5 displays the set of passes generated from a request. Note that the duration of these passes are between the minimum and maximum specified by the request, and inside the time window defined by the release and due times.

3.1.3.4 Redundancy

Satellites and ground stations generally have limited communication capacity, and in fact SRS literature generally considers that each satellite may only communicate to a ground station at a time, and vice-versa. In this subsection we will present those constraints to finally provide a definition of feasible schedule.

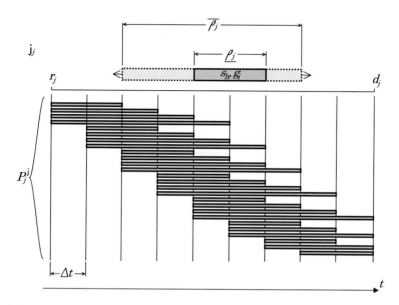

Fig. 3.5 Transformation of a request into a set of passes, with all the possible combinations of start times and durations complying with the specification of the request, and taking into account the duration of the discretization step

For the sake of clarity, for each $p_k \in P_{\text{sub}} \subseteq P = \{P_j^i\}$, the functions ϕ_g and ϕ_s are defined (e.g., $\phi_g : p_k \longrightarrow g_i : p_k = (s_h, g_i, n_{s_k}, n_{e_k}, w_k))$ for accessing the applicable elements in the tuple (so that $\phi_g(p_k) = g_i$ and $\phi_s(p_k) = s_h$), and the inverse D_n^{-1} for accessing the request originating the pass ($D_n^{-1} : p_k \longrightarrow j_j : p_k \in D_n(j_j)$). Note that this check is necessary for avoiding conflicts in passes generated from the same request. Allowing these conflicts is known in the literature as *preemption*, which as stated previously will not be considered in SRS.

Let C_g and C_s be conflict indicator boolean functions which yield a 1 if two passes $p_u, p_v \in P_{\text{sub}} : u, v \in \mathbb{N} \cap [1, |P_{\text{sub}}|]$ are generated by the same request or overlapping in time for a single ground station (C_g) or satellite (C_s):

$$C_g : p_u, p_v \longrightarrow \begin{cases} 1, & \text{if } [D_n^{-1}]_{u,v} \vee \left([n]_{u,v} \wedge [g]_{u,v}\right), \\ 0, & \text{otherwise}, \end{cases} \quad (3.1.17)$$

$$C_s : p_u, p_v \longrightarrow \begin{cases} 1, & \text{if } [D_n^{-1}]_{u,v} \vee \left([n]_{u,v} \wedge [s]_{u,v}\right), \\ 0, & \text{otherwise}, \end{cases} \quad (3.1.18)$$

where the bracketed boolean variables

$$[D_n^{-1}]_{u,v} = 1 \Leftrightarrow D_n^{-1}(p_u) = D_n^{-1}(p_v), \quad (3.1.19)$$

$$[g]_{u,v} = 1 \Leftrightarrow \phi_g(p_u) = \phi_g(p_v), \quad (3.1.20)$$

$$[s]_{u,v} = 1 \Leftrightarrow \phi_s(p_u) = \phi_s(p_v), \tag{3.1.21}$$

$$[n]_{u,v} = 1 \Leftrightarrow \{(n_{s_v} \in [n_{s_u}, n_{e_u}]) \vee (n_{e_v} \in [n_{s_u}, n_{e_u}])\}, \tag{3.1.22}$$

identify the types of conflicts (same request, same ground station, same satellite, or time overlapping).

Let $p_k \times P_{sub}$ be the Cartesian product of the element p_k with the set P_{sub}, so that $p_k \times P_{sub} = \{p_k, p_v\} : p_k, p_v \in P_{sub} \ \forall k, v \in \mathbb{N} \cap [1, |P_{sub}|]$. Now let C_G and C_S be the functions which generate the number of conflicts of a schedule P_{sub}, only for ground stations in the first case and only for satellites in the second:

$$C_G : p_k, P_{sub} \longrightarrow \sum C_g(p_k \times P_{sub}) = \sum_{v=1}^{|P_{sub}|} C_g(p_k, p_v) : k \neq v, \tag{3.1.23}$$

$$C_S : p_k, P_{sub} \longrightarrow \sum C_s(p_k \times P_{sub}) = \sum_{v=1}^{|P_{sub}|} C_s(p_k, p_v) : k \neq v. \tag{3.1.24}$$

Combining the functions C_G and C_S, and simplifying the notation, let C_Σ be the function which generates the total number of conflicts of a schedule P_{sub}, summing the number of conflicts for ground stations and satellites. Thus, for every $p_k \in P_{sub}$:

$$C_\Sigma : P_{sub} \longrightarrow \sum C_G(P_{sub}) + \sum C_S(P_{sub})$$
$$= \sum_k C_G(p_k) + \sum_k C_S(p_k). \tag{3.1.25}$$

Figure 3.6 shows a subset of passes generated from requests j_1, j_2, and j_3. Passes p_1, p_2, and p_5 are associated to j_1, passes p_3 and p_6 to j_2, and pass p_4 to j_3. From the figure and according to Eqs. (3.1.17)–(3.1.25), it is easy to see that the pass p_1 is conflicting with p_2, p_3 and p_5; p_2 with p_1, p_3 and p_5; p_3 with p_1, p_2 and p_6; p_4 does not conflict with any other pass in this subset; p_5 is conflicting with p_1, p_2 and p_6; and p_6 with p_3 and p_5.

Definition 5. A *nonredundant* schedule must be free of both kinds of conflicts (C_G and C_S). This case is also identified as *unitary capacity*. Alternatively, a certain number of conflicts of a kind are allowed in the *redundant* schedule.

$$\text{Unitary capacity} \Leftrightarrow C_\Sigma(P_{sub}) = 0, \tag{3.1.26}$$

$$m\text{-ary capacity} \Leftrightarrow \begin{cases} C_G(p_i, P_{sub}) \leq m - 1 \ \forall p_i \in P_{sub}, \\ C_S(P_{sub}) = 0, \end{cases} \tag{3.1.27}$$

where C_G and C_S can be exchanged depending on the kind of entities with m-ary capacity. It is easy to see that unitary capacity is a subproblem of m-ary capacity, as it is the case where $m = 1$.

Note that if m is maximal, that is, if no conflicts are checked for the satellites (or ground stations), the problem can be split into several one ground station (or satellite) problems. Unitary capacity is the case generally considered in SRS

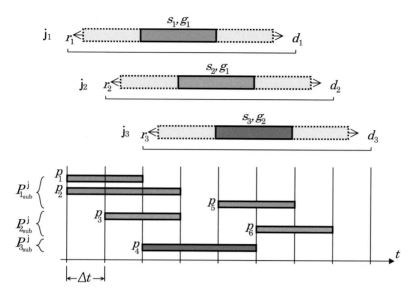

Fig. 3.6 Examples of conflicting passes

literature [1–3], although some references do consider the multiple capacity case [4]. In this book we also focus on the unitary capacity case, although we also provide an optimal solution for the problem with multiple capacity in Chap. 4 (§4.3.2).

3.1.3.5 Precedence

Precedence constraints allow us to model interdependencies among requests.

Definition 6. A problem has *precedence* constraints if its definition includes sets P_l^P (with $l = 1, 2, \dots$) of passes which cannot be on the schedule without each of their elements. Also, by definition, these passes will have end times smaller than the start time of the pass. Otherwise there are *no-precedence* constraints.

$$\text{No-precedence} \Leftrightarrow \nexists P^P : \forall p_k \in P^P \subset P, \tag{3.1.28}$$
$$p_k \in P_{\text{sub}} \Leftrightarrow P^P \subset P_{\text{sub}},$$

$$\text{Precedence} \Leftrightarrow \exists P_l^P : \forall p_k \in P_l^P \subset P, \tag{3.1.29}$$
$$p_k \in P_{\text{sub}} \Leftrightarrow P_l^P \subset P_{\text{sub}}.$$

Precedence constraints are generally external to the problem and defined on the requests, however, it is easy to propagate them in Transformation 1 to generate the subsets P_l^P. By definition no-precedence is a subproblem of precedence, although most of the literature considers no-precedence.

Note that we consider precedence sets to be either completely included or dismissed from the schedule as in [3], contrasting to the approach generally taken in general scheduling [12].

Definition 7. A *feasible schedule* P^{f} is a nonpreemptive schedule that meets the constraints for redundancy (redundant or nonredundant) and precedence (precedence constraints existing or not) as specified for the problem.

3.1.3.6   Priority

Priorities allow us to model the preference among requests.

Definition 8. The weight or priority w_k introduced in the discretization (3.1.9) will depend on the weight function of the originating request $f_j^{\text{w}}(t)$ (introduced in [1] as the *suitability function*) for each pass, and on the limits of the new window (n_{s_k}, n_{e_k}). Furthermore a normalization factor a_k is introduced:

$$w_k \triangleq \sum_{n=n_{s_k}}^{n_{e_k}} a_k f_j^{\text{w}}(n\Delta t). \tag{3.1.30}$$

Prioritization approaches depend on the values of a_k and $f_j^{\text{w}}(n\Delta t)$. Let $\Delta n_k = n_{e_k} - n_{s_k}$ and $\Delta n_{\max} = \max(\Delta n_k)$ $\forall k$ such that $p_k \in P_j$. A problem is said to have *no-priority* constraints if all the requests are assigned the same priority, which for simplicity is 1 $(a_k = \Delta n_k^{-1}; f_j^{\text{w}}(n\Delta t) = 1 \Rightarrow w_k = 1)$. The general case is said to have *priority* constraints, and takes into account both duration and best spot of the pass $(a_k = \Delta n_{\max}^{-1}; f_j^{\text{w}}(n\Delta t) = f_j^{\text{suit}}(n\Delta t) \Rightarrow w_k = \Delta n_{\max}^{-1} \sum_{n=n_{s_k}}^{n_{e_k}} f_j^{\text{suit}}(n\Delta t))$.

It is easy to see that nonprioritized problems are subproblems of prioritized problems.

An example of the generation of the priorities associated to the passes can be seen in Fig. 3.7, with two passes of durations: $\Delta n_1 = 4$ and $\Delta n_2 = 12$. Let us assume that $\Delta n_{\max} = \Delta n_2$, then the pass that is shorter and is on a time range where the suitability function ranks low has approximately a priority $w_1 = 0.1$, whereas the pass that has a longer duration and is in the time region where the function ranks higher has an approximate priority $w_2 = 0.9$. Note that we are considering the general case presented in Definition 8.

3.1.4   Schedule Metrics

In order to characterize the quality of the schedules it is necessary to introduce a metric, which finally allows to define the SRS problem.

Definition 9. Given a schedule P_{sub}, let the metric $\| \cdot \|_{\Sigma w}$ be:

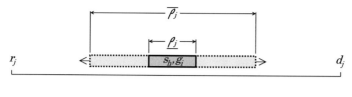

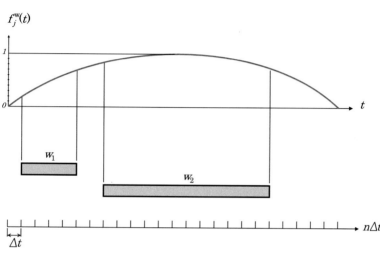

Fig. 3.7 Priorities of the passes generated from a request

$$\|P_{\text{sub}}\|_{\Sigma w} \triangleq \sum_k w_k \ \forall p_k \in P_{\text{sub}} . \tag{3.1.31}$$

Finally, the SRS problem can be defined as finding the optimal schedule.

Definition 10. Given an initial set of passes P, and the set of all the feasible schedules $\{P^f\}$, the *Satellite Range Scheduling (SRS) problem* can be stated as finding the optimal schedule P^*:

$$P^* \triangleq \arg \max(\|P_{\text{sub}}\|_{\Sigma w}) \ \forall P_{\text{sub}} \in \{P^f\}, \tag{3.1.32}$$

or equivalently:

$$P^* \in \{P^f\}; \ \nexists P_{\text{sub}} \in \{P^f\} \ : \|P_{\text{sub}}\|_{\Sigma w} > \|P^*\|_{\Sigma w}, \tag{3.1.33}$$

where $\{P^f\}$ is the set of all the feasible schedules.

Some references [2, 4, 5] add to the problem the term *oversubscribed*, when not all the requests can be served ($|P^*| \neq |J|$).

3.2 Complexity of SRS

The purpose of this section is to classify the main SRS problems in terms of complexity. Whereas complexity for specific cases has been studied in some references [1–3, 5, 7], most of the literature on SRS focus on suboptimal algorithms.

3.2.1 Introduction to Complexity Theory

We provide a very short introduction to complexity theory, for a deeper insight reading of [13] is encouraged. This introduction should however be enough for providing those readers less experienced in computational complexity theory with a basic framework for supporting all the complexity results presented in this book.

The SRS problem is an *optimization problem* [finding the schedule with maximal metric (3.1.32)], and as such, it shares similar structure with its *decision problem* version (determine whether a schedule with a metric higher than a certain value exists or not). Optimization problems are at least as hard as their decision versions [13], so let us focus for now on the latter. Those problems for which a polynomial time algorithm is known are said to be in the *Class P* (polynomial). There is a more general class of problems, the *Class NP* (nondeterministic polynomial) (P ⊆ NP), for which their solution can be verified in polynomial time. Problems in P are polynomially equivalent (a polynomial transformation between them exists), and there is a sub-class in NP that shares this property, which is the class of *NP-complete* problems. Those problems that are at least as complicated as NP-complete problems (i.e., they can be transformed into these), but are not necessarily in NP are said to be *NP-hard*. Getting back to optimization problems, if a decision version is NP-complete, then the optimization problem is NP-hard [13]. It is generally considered that if a problem is not in P then it is *intractable*.

Given a certain problem, a subproblem of it asks the same question but over a reduced set of instances. This does not mean that having additional constraints will generate a subproblem, since these constraints may rise the size of the input of the problem, thus allowing for more instances. A subproblem of a problem in P is also in P, and a problem is NP-hard if it has a subproblem that is also NP-hard [13].

The computational complexity of the algorithms presented in this book will be given in big O notation. That is, we say that an algorithm runs in $O(f(N))$, where $f(N)$ is a function of the size of the input of the problem, if the execution time of the algorithm can be bounded by the product of a constant and this function.

3.2.2 Complexity of the SRS Problem

We now provide some results for classifying the main SRS problems.

Lemma 3.1. *SRS is NP-hard.*

Proof. Barbulescu et al. [2] prove equivalence of the single resource satellite scheduling problem with fixed-slack, no-preemption, no-redundancy, no-precedence, and no-priorities to the general scheduling problem of the minimization of the number of late jobs in one machine where tasks have different release times. This last problem is classified as NP-hard in [3, 14].

From the reducibility relations given in Definitions 2, 5, 6, and 8, the generalized problems with variable-slack, multiple resources, redundancy, precedence, and priorities have to be at least as complicated as the subproblem, so that they are also NP-hard.   □

Theorem 3.1. *Deciding discrete no-slack SRS is NP-complete for arbitrary numbers of ground stations and satellites.*

Proof. The aim of this demonstration is to show equivalence of this problem to an existing NP-complete problem. We start with the subproblem no-slack MuRRSP with no-redundancy, no-preemption, no-precedence, and no-priority constraints. A graph will be created from the initial set of passes P. Every pass will be mapped into a node in the graph (due to the no-slack condition $|P| = |J| = N$ nodes). Undirected edges will be created between every two nodes if they are not conflicting $C_g(p_u, p_v) = 0$ and $C_s(p_u, p_v) = 0$ for every $p_u, p_v \in P$. Then finding a feasible schedule of size M is equivalent to finding a clique of size M in the undirected graph (known as the "Clique" problem [13]), which decision version is NP-complete.

Finding the size of the optimal schedule, from Definition 10, is equivalent to solving this decision problem at most N times (for $M = 1, 2, \ldots, N$), and then selecting the solution with the highest metric (which for the nonprioritized case is the number of passes). Then if Algorithm A solves the decision problem in $O(f_A(N))$ time (where $f_A(N)$ is a function of N), the maximal size problem can be solved in $O(N f_A(N))$ time, and thus it is also NP-complete. Thus, the no-slack MuRRSP decision problem with no-redundancy, no-preemption, no-precedence, and no-priority is NP-complete, and its optimization version will be NP-hard.

From the reducibility relations given in Definitions 5, 6, and 8, the generalized problems with redundancy, precedence, and priorities have to be at least as complicated as the subproblem, so that they are also NP-hard.   □

The problem will still be intractable if we consider precedence constraints.

Corollary 3.1. *SRS with precedence constraints is NP-hard, even for a single ground station or satellite, no-slack and no-priorities.*

Proof. Let us suppose a scenario with a single scheduling entity with no-slack, no-preemption, no-priorities, but with precedence constraints. We generate a graph similarly as in Theorem 3.1, where nodes represent passes (and precedence sets) and their priority (or sum of priorities for the precedence sets), and in this case edges are created between pairs of nodes if their associated passes (or precedence sets) are conflicting. These graphs are also known in the literature as multiple-interval graphs, or t-interval graphs, where t is the maximum cardinality of the precedence sets. It is easy to see that finding the optimal schedule in this case is equivalent to finding the maximum-weight independent set, that is, the set of nodes of the graph where no pair of nodes is connected by an edge, and which sum of associated weights is maximum. Based on the properties of the created graph, this set of nodes corresponds to the feasible (independent set) schedule with maximum metric (maximum-weight).

Reference [15] shows that the maximum independent set problem is NP-hard for t-interval graphs with $t \geq 2$. Given that it is a subproblem of the maximum-weight independent set (where the weights are unitary), the maximum-weight problem is also NP-hard, and therefore single resource satellite range scheduling with no-slack, no-preemption, no-priorities, but with precedence is NP-hard. This result is thus extensible to the multiple entities, slack, prioritized versions of the problem.   □

The problem is simplified though if we remove the precedence constraints, and have a fixed number of ground stations or satellites:

Theorem 3.2. *Discrete no-slack no-precedence SRS is in class P for a fixed number of ground stations or satellites.*

Proof. This problem will be shown to be a subproblem of a problem known to be solvable by a polynomial time algorithm. Suppose a discrete no-slack MuRRSP problem with unitary capacity, priorities, and no-precedence. This problem corresponds to a subproblem of the general scheduling problem defined in [14], where machines can be associated to either ground stations or satellites, and the complementary defines additional conflicts, thus reducing the space of feasible states of the system. In the SRS version every request is assigned to a single "machine" due to the structure of the requests (3.1.2), which makes the problems easier as it reduces the number of states in the graph.

Given that the problem from [14] can be solved in polynomial time for a fixed number of machines, the result is also extensible to the no-slack no-redundant no-precedence SRS problem with a fixed number of ground stations or satellites. From Definitions 2 and 8, SiRRSP is a subproblem of MuRRSP, and no-priority of priority. For the m-ary capacity case, we can transform those passes associated to multiple capacity entities into m passes associated to m fictitious entities in polynomial time, thus generating an instance of a problem with unitary capacity. Therefore, the result is also extensible to the rest of the cases, completing the proof for discretized no-slack no-precedence SRS.   □

Some tedious *manipulation in the notation allows to extend the results from Theorems 3.1 and 3.2 to the continuous time versions of the problems*, as there is no difference other than taking discrete values for the times when checking for the conflicts.

The two following results are however constrained to the discretized versions of the problem, as they take advantage of a polynomial transformation (3.1.9) only available in the discretized problem.

Corollary 3.2. *Deciding discretized SRS is NP-Complete for arbitrary numbers of ground stations and satellites.*

Proof. From Proposition 3.1, the equivalent set of passes can be generated in polynomial time from the set of requests with discrete time. Then following the procedure from Theorem 3.1 a similar graph can be generated, and following the same reasoning the problem is also NP-complete. □

Corollary 3.3. *Discretized no-precedence SRS is in class P for a fixed number of ground stations and satellites.*

Proof. From Proposition 3.1, the discretized slack problem is polynomially transformable to the discretized no-slack problem in polynomial time. Note that passes generated from the same request may be conflicting even though not time overlapping, and given that there may be $|G||S|$ requests time overlapping at the same time, we consider in this case both the number of ground stations and satellites to be fixed in order to bound the number of schedules between two passes generated from the same request. Applying Theorem 3.2 yields that discretized no-precedence SRS with a fixed number of ground stations and satellites is in polynomial class, since the discretized no-slack version is as well. □

Note that Theorem 3.1 and Corollary 3.2 refer to decision versions of the problem, so that the optimization versions are NP-hard.

Therefore, the satellite range scheduling problem can be solved in polynomial time if time is discretized, the number of satellites and ground stations is fixed and there are no-precedence constraints.

3.3 General Scheduling Problems

We have provided the connection between multiple entity general machine scheduling and satellite range scheduling in the previous section (Theorem 3.2). In this section we will provide further insight on the connections on more variants of the problem.

We will apply an existing widely used notation from general scheduling (presented in [12]), allowing to formally relate problems from the two scheduling fields. This notation is summarized for the sake of completeness (and extended to cover SRS specific cases), and a survey is presented to shed light on existing and new relations between specific problems.

3.3.1 Problem Classification

In general scheduling problems [12] there is a set of n jobs $\{\mathfrak{J}_j\} \forall j \in \mathbb{N} \cap [1, n]$ to be processed in a set of m machines $\{\mathfrak{M}_i\} \forall i \in \mathbb{N} \cap [1, m]$. Furthermore, each job $\mathfrak{J}_j$ has a processing time $\mathfrak{p}_j$ ($\mathfrak{p}_{ij}$ if it is different for each machine $\mathfrak{M}_i$), a release date

$\mathfrak{r}_j$, a due date $\mathfrak{d}_j$, a weight (or priority) $\mathfrak{w}_j$ and a cost function $\mathfrak{f}_j(t)$ for evaluating the penalization of non satisfying the request before t.

Note that this notation is similar to the previously introduced, with the following differences: n is used for the number of requests N, the set of m machines will be $|S|$ or $|G|$ depending on whether the scheduling entities are the satellites or the ground stations, and the time duration $\mathfrak{p}$ substitutes ρ and its discrete homologous.

Graham et al. [12] defines general scheduling problems based on three parameters: $\alpha|\beta|\gamma$. The possible values for these parameters are summarized in Fig. 3.8, where newly introduced notation is marked with the symbol *.

α: Relation Among the Machines The first parameter α designates the relation among the machines: 1 for a single machine; $\mathfrak{P}$ for parallel identical machines (processing time for a task is the same for every machine); $\mathfrak{R}$ for unrelated parallel machines (processing times are different and even a task can be only compatible with a subset of machines).

β: Constraints The second parameter β covers the constraints on the jobs. Terms grouped in the same bullet are mutually exclusive:

Release times   The term $\mathfrak{r}_j$ indicates different release times for every job, which is generally assumed for SRS, due to the association among requests and visibility windows.

Durations   We introduce $\overline{\overline{\mathfrak{p}}}_j$, for indicating that the processing time extends from the release date to the due date (no-slack case), $\overline{\overline{\mathfrak{p}}}_j = \mathfrak{d}_j - \mathfrak{r}_j$; and another notation is introduced (for the variable-slack case), $\underline{\mathfrak{p}}_j \leqslant \mathfrak{p}_j \leqslant \overline{\mathfrak{p}}_j$ (existing notation $\underline{\mathfrak{p}} \leqslant \mathfrak{p}_j \leqslant \overline{\mathfrak{p}}$ assumes constant bounds for all the jobs) for indicating that the processing times are allowed to be within lower and upper bounds. The absence of these constraints indicates that $\mathfrak{p}_j$ is fixed and $0 \leqslant \mathfrak{p}_j \leqslant \overline{\overline{\mathfrak{p}}}_j$ (fixed-slack case), which is the general case in general scheduling problems.

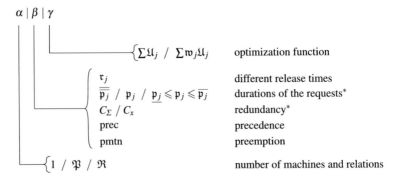

Fig. 3.8 Unified notation for representing scheduling problems

Conflicts   We introduce C_Σ, which is specific for SRS, and indicates that no-redundancy is allowed in tasks assignable to an only machine (whether station or satellite). As noted in Definition 5, it is indifferent whether to apply it or not in SiRRSP. We also introduce C_x to represent the case where the number of conflicts either for satellites or ground stations is reduced to a certain number.

Precedence   The term "prec" introduces precedence constraints, for cases where some tasks require the completion of others to be served.

Preemption   The term "pmtn" allows for preemption, which is not considered for SRS, and thus will not be used in any case.

γ: Optimization Function The third parameter γ describes the optimization function: $\sum \mathfrak{U}_j$ stands for the total number of late jobs ($\mathfrak{U}_j = 1$ if a job $\mathfrak{J}_j$ is late); and $\sum \mathfrak{w}_j \mathfrak{U}_j$ takes also into account the weights assigned to each job in the computation. Minimizing the sum of the weight of late jobs is equivalent to maximizing the sum of the weight of the jobs in the schedule.

3.3.2 Problem Reducibility

Graham et al. [12] also describes the reducibility between the general scheduling problems. In this subsection equivalent relations will be obtained combining those from the reference with the definitions from §3.1.

The variable-slack problem $\underline{p_{ij}} \leqslant p_{ij} \leqslant \overline{p_{ij}}$ can be seen as a generalization of the fixed-slack problem $\overline{\overline{p_j}}$ (also known as knapsack problem [1]). And given that $\overline{\overline{p_j}}$ is a specific case of p_{ij} (when $p_{ij} = \overline{\overline{p_j}} \ \forall j$), and p_j is a specific case of $\underline{p_{ij}} \leqslant p_{ij} \leqslant \overline{p_{ij}}$ (when $\underline{p_{ij}} = \overline{p_{ij}}$):

$$1 \subset \mathfrak{R}, \tag{3.3.1}$$

$$\sum \mathfrak{U}_j \subset \sum \mathfrak{w}_j \mathfrak{U}_j, \tag{3.3.2}$$

$$\overline{\overline{p_j}} \subset p_{ij} \subset \underline{p_{ij}} \leqslant p_{ij} \leqslant \overline{p_{ij}}. \tag{3.3.3}$$

This reference also shows that no-precedence is a subproblem of precedence, but note that the approach presented in [12] for precedence is more general than that in Definition 6. Nevertheless, we have shown that even the simplest problem with precedence in SRS is NP-hard (Corollary 3.1).

Some of the relations between general and satellite-specific problems will be summarized in the following subsections. Note that from the previous relations, properties, and references are inherited by the problem subsets, so that references will not be repeated unless explicitly necessary.

3.4 Relating Satellite and General Scheduling Problems

In this section the relations between general scheduling (GS) and satellite scheduling (SRS) problems are explored. As usual, *no-preemption is considered for SRS*.

3.4.1 One Machine Problems

One machine scheduling problems correspond to different variants of SiRRSP. Note that this constraint on the number of machines eliminates the relevance for the redundancy definitions, so both versions will be equivalent.

3.4.1.1 $1 \mid r_j, \overline{\overline{p_{ij}}} \mid \sum \mathcal{U}_j$

We start with the simplest case, that is, a single machine, no-slack, no-precedence, and no-priorities on the requests:

GS Arkin and Silverberg [14] provide an optimal solution in $O(n^2)$ as a specific case of a more complex problem.

SRS Burrowbridge [7] provides a demonstration of the equivalence of this problem with the *no-slack no-precedence no-priorities SiRRSP*, as well as an algorithm for finding the *optimal* solution in polynomial time (greedy earliest deadline).

3.4.1.2 $1 \mid r_j, \overline{\overline{p_{ij}}}, \text{prec} \mid \sum \mathcal{U}_j$

Compared to the previous case we introduce precedence constraints:

GS Scheduling for a single machine where requests have fixed intervals with precedence constraints (as defined in Definition 6) is known in the literature as multiple-interval or t-interval scheduling. Fellows et al. [15] shows NP-hardness of this problem for $t \geqslant 2$.

SRS A demonstration of the equivalence of this problem to the *no-slack no-prioritized SiRRSP with precedence constraints* has been provided in Theorem 3.1 in this chapter.

3.4.1.3 $1 \mid r_j, \overline{\overline{p_{ij}}} \mid \sum w_j \mathcal{U}_j$

If the requests have priorities associated to them, then the greedy algorithm with the earliest deadline heuristic is no longer valid (see §4.4 for an example). There are algorithms for machine scheduling also valid for the SRS version of the problem, based on the results previously presented in this chapter:

GS   An optimal solution in $O(n^2)$ is provided in [14], based on finding a shortest path in a directed acyclic graph.

SRS   A demonstration of the equivalence of this problem to the *no-slack no-precedence prioritized SiRRSP* has been provided in Theorem 3.2 in this book. The optimal solution is presented in Chap. 4 in this book for a generalization of this problem, and we provide an optimal $O(n)$ algorithm based on an algorithm from a variant of SRS in Chap. 6 (§6.5).

3.4.1.4 $1 \mid r_j \mid \sum w_j \mathfrak{U}_j$

Compared to the previous problem, in this case the requests have fixed-slack. This is the problem historically addressed in previous SRS literature. Its complexity is well known for machine scheduling, and further results on the SRS version of this problem have been provided in this chapter for the next problem of the list.

GS   Graham et al. [12] show NP-hardness of this problem.

SRS   Barbulescu et al. [2] show equivalence of this problem to *fixed-slack no-precedence prioritized SiRRSP*.

3.4.1.5 $1 \mid r_j, \underline{p_{ij}} \leqslant p_{ij} \leqslant \overline{p_{ij}} \mid \sum w_j \mathfrak{U}_j$

This is the general case for a single scheduling entity: variable-slack, no-precedence, and priorities.

GS   From the reducibility relation (3.3.3) (see also [12]), this problem is at least as complex as the previous one, which is NP-hard.

SRS   This problem is specified in [1], and it is by definition equivalent to the *variable-slack no-precedence prioritized SiRRSP*. From Corollary 3.3 in this book (§3.2) the discretized version of this problem can be solved in polynomial time. The optimal solution is presented in Chap. 4 (§4.3.1) in this book for a generalization of this problem with discretized time.

3.4.2   Several Identical Machines Problems

These problems correspond to the case of scheduling in an entity with limited capacities, which for simplicity can be broken down into basic entities with unitary capacity. Specifically, for satellite scheduling, these problems are equivalent to the scenarios where it is possible to find a subset division of the set of requests, such that all the subsets have the same elements:

$$Q = \{Q_l\} \ : \ Q_l = Q'_l \ \forall l, l'. \tag{3.4.1}$$

An illustrative example of this scenario is scheduling for a ground station with different independent and compatible communication systems, as they share the same position and thus the scheduling tasks can be assigned to each of them. The usefulness of this approach, specially for the case of LEO (Low Earth Orbit) tracking stations, is limited to scenarios where the coverage regions do not overlap.

By definition, these problems are not equivalent to any of the ones that fit on the notation used in this text.

3.4.2.1 $\mathfrak{P} \mid r_j, \overline{\overline{p_{ij}}}, C_\Sigma \mid \sum \mathfrak{U}_j$

We only consider a problem previously tackled in the SRS literature for reference, since more general problems are treated in the next subsection. This case corresponds to the extension of the first problem detailed in the previous subsection for a single machine, but considering several identical machines instead of one.

GS   Arkin and Silverberg [14] provide an optimal solution in $O(n^2\log(n))$ as a specific case of a more complex problem. Also several references are provided in the survey on interval scheduling in [16].

SRS   Barbulescu et al. [2] provide an optimal solution for this problem in polynomial time. Note that in SRS combining the constraint C_Σ with similar resources and passes assigned to an only resource translates duplicate passes (except for the assigned resource) into single tasks assignable to any machine.

3.4.3   *Several Unrelated Machines Problems*

As introduced before, problems involving several unrelated machines are suitable to be applied to different variants of MuRRSP.

3.4.3.1 $\mathfrak{R} \mid r_j, \overline{\overline{p_{ij}}}, C_\Sigma \mid \sum w_j \mathfrak{U}_j$

Again we start with the most simple version of the problem, which is basically the extension of the third case in the list of single machine scheduling, but with different machines instead of one. Based on the transformation presented in this chapter, this version of the problem will be key for tackling others.

GS   N/A (not applicable). Due to the constraint C_Σ this problem is specific to the satellite scheduling domain.

SRS   A demonstration of the equivalence of this problem to the *no-slack no-redundant no-precedence prioritized MuRRSP* has been provided in Theorem 3.2 in this book. Also it is proved that the problem is in class P for a fixed number of machines. The optimal solution is presented in Chap. 4 (§4.2) in this book.

3.4.3.2 $\mathfrak{R} \mid \mathfrak{r}_j, \overline{\overline{\mathfrak{p}_{ij}}}, C_x \mid \sum \mathfrak{w}_j \mathfrak{U}_j$

Compared to the previous case we introduce redundancy.

GS   N/A.

SRS   This problem is specified in [4], and it is by definition equivalent to the *no-slack no-precedence prioritized SiRRSP with redundancy*. We showed in Theorem 3.2 in this book that this problem can be solved in polynomial time for a fixed number of entities, and we present in Chap. 4 (§4.3.2) an optimal solution to this problem.

3.4.3.3 $\mathfrak{R} \mid \mathfrak{r}_j, \overline{\overline{\mathfrak{p}_{ij}}}, C_\Sigma, \mathbf{prec} \mid \sum \mathfrak{w}_j \mathfrak{U}_j$

Compared to the first case presented for several unrelated machines, in this precedence is considered. This version of the problem is introduced here since a more complex one has been approached in existing SRS literature, to be treated later.

GS   N/A.

SRS   A demonstration of the equivalence of the subproblem with a single machine and no-priorities to the *no-slack no-prioritized SiRRSP with precedence constraints* problem has been provided in Theorem 3.1 in this chapter. Since the subproblem is NP-hard, so it is this problem.

3.4.3.4 $\mathfrak{R} \mid \mathfrak{r}_j, C_\Sigma \mid \sum \mathfrak{w}_j \mathfrak{U}_j$

This problem, as it was its single machine version, is one of the most studied problems in SRS literature. Note the additional constraint in the SRS problem compared to machine scheduling. As a subproblem of the last one in this list, the results provided in this chapter for the last one are applicable to this version too.

GS   N/A.

SRS   Equivalence to *fixed-slack no-redundant no-precedence prioritized MuRRS* of this problem, as well as NP-hardness, are demonstrated in [3]. A survey on several suboptimal algorithms is provided in [5] for the discretized version of the problem.

3.4.3.5 $\mathfrak{R} \mid \mathfrak{r}_j, C_\Sigma, \mathbf{prec} \mid \sum \mathfrak{w}_j \mathfrak{U}_j$

This problem adds precedence to the previous one. It has been included in this list given that it has been tackled in previous SRS literature.

GS   N/A.
SRS   Equivalence to *fixed-slack no-redundant prioritized MuRRS with precedence constraints* of this problem, as well as NP-hardness, are demonstrated in [3]. Additional constraints are applied, specifically setup times (which could be modeled on the suitability function and start time) and on-board data storage limits.

3.4.3.6 $\mathfrak{R} \mid \mathfrak{r}_j, \underline{\mathfrak{p}_{ij}} \leqslant \mathfrak{p}_{ij} \leqslant \overline{\mathfrak{p}_{ij}}, C_{\Sigma} \mid \sum \mathfrak{w}_j \mathfrak{U}_j$

This is the most general problem in this list: several machines, variable-slack, no-redundancy, and priorities. Results have been provided in this chapter regarding the complexity of this problem and the solution of its discretized version. Also an algorithm is provided for this problem with discretized time in the next chapter, which is thus applicable to its subproblems.

GS   N/A.
SRS   This problem is defined in [1], and it is by definition equivalent to *variable-slack no-redundant no-precedence prioritized MuRRSP*. As a generalization of the analogous fixed-slack problem, it is at least NP-hard. From Corollary 3.3 in this book, the discretized version of this problem can be solved in polynomial time for a fixed number of ground stations and satellites. The optimal solution for this discretized version is presented in Chap. 4 (§4.3.1) in this book.

3.5   Summary

Characteristics of the enumerated problems have been tabulated into Table 3.1 for easy consultation. Although a $\mathfrak{P}$ (several identical machines) problem has been included in this table as MuRRSP (marked as x'), it is a very specific case (all the resources have the same position), so as stated previously, the applicability of this case is found to be reduced.

The notation $\underline{\mathfrak{p}_{ij}} \leqslant \mathfrak{p}_{ij} \leqslant \overline{\mathfrak{p}_{ij}}$ has been changed to $\mathfrak{p}_{ij}^{\mathrm{var}}$, and some terms have been abbreviated: Res. (resource), Si. (single), Mu. (multiple), Prec. (precedence), Red. (redundancy), Fix. (fixed), and Var. (variable). The complexity of these problems is displayed for the two extreme cases studied in §3.2: t (general case in continuous time), and Δt,F (discretized time case with fixed number of entities). The classes of complexity are H (for NP-hard) and P (for polynomial).

Figure 3.9 summarizes the complexity and relations among the surveyed SRS problems, with arrows identifying subproblems (origins) and generalizations (ends). Note that optimal solutions for SRS problems prior to this work reduce to the problem $1 \mid \mathfrak{r}_j, \overline{\overline{\mathfrak{p}_{ij}}} \mid \sum \mathfrak{U}_j$ represented at the bottom of the graph, solved in [7], and to $\mathfrak{P} \mid \mathfrak{r}_j, \overline{\overline{\mathfrak{p}_{ij}}}, C_{\Sigma} \mid \sum \mathfrak{U}_j$, solved in [2] as a generalization of the previous one. We

Table 3.1 Complexity of general and satellite scheduling. [9] (extended) ©2014 Springer

Problem	Res.		Priority		Prec.		Red.		Slack			Class		References	
	Si.	Mu.	No	Yes	No	Yes	No	Yes	No	Fix.	Var.	t	$\Delta t,F$	GS	SRS
$1|\tau_j,\overline{\overline{p_{ij}}}|\sum \mathfrak{U}_j$	x		x		x		x		x			P	P	[14]	[7]
$1|\tau_j,\overline{\overline{p_{ij}}},\text{prec}|\sum \mathfrak{U}_j$	x		x			x	x		x			H	H	[15]	§3
$1|\tau_j,\overline{\overline{p_{ij}}}|\sum w_j\mathfrak{U}_j$	x			x	x		x		x			P	P	[14]	§3, §6
$1|\tau_j|\sum w_j\mathfrak{U}_j$	x			x	x		x			x		H	P	[12]	[2]
$1|\tau_j,p_{ij}^{\mathrm{var}}|\sum w_j\mathfrak{U}_j$	x			x	x		x				x	H	P	[12]	[1],§3,§4
$\mathfrak{R}|\tau_j,\overline{\overline{p_{ij}}},C_\Sigma|\sum \mathfrak{U}_j$		x'	x		x		x		x			P	P	[14, 16]	[2]
$\mathfrak{R}|\tau_j,\overline{\overline{p_{ij}}},C_\Sigma|\sum w_j\mathfrak{U}_j$		x		x	x		x		x			H	P	N/A	§3,§4
$\mathfrak{R}|\tau_j,\overline{\overline{p_{ij}}},C_x|\sum w_j\mathfrak{U}_j$		x		x	x			x	x			H	P	N/A	[4],§3,§4
$\mathfrak{R}|\tau_j,\overline{\overline{p_{ij}}},C_\Sigma,\text{prec}|\sum w_j\mathfrak{U}_j$		x		x		x	x		x			H	H	N/A	§3
$\mathfrak{R}|\tau_j,C_\Sigma|\sum w_j\mathfrak{U}_j$		x		x	x		x			x		H	P	N/A	[3, 5]
$\mathfrak{R}|\tau_j,C_\Sigma,\text{prec}|\sum w_j\mathfrak{U}_j$		x		x		x	x			x		H	H	N/A	[3],§3
$\mathfrak{R}|\tau_j,p_{ij}^{\mathrm{var}},C_\Sigma|\sum w_j\mathfrak{U}_j$		x		x	x		x				x	H	P	N/A	[1],§3,§4

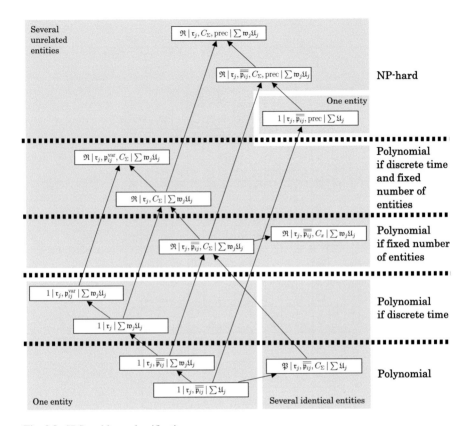

Fig. 3.9 SRS problems classification

have shown that SRS is a subproblem of general scheduling for unrelated machines where each request is assigned to a pair of resources from disjoint sets.

Acknowledgements This research was performed while the author held a National Research Council Research Associateship Award at the Air Force Research Laboratory (AFRL).

References

1. W.J. Wolfe, S.E. Sorensen, Three scheduling algorithms applied to the Earth observing systems domain. Manag. Sci. **46**(1), 148–168 (2000)
2. L. Barbulescu, J.P. Watson, L.D. Whitley, A.E. Howe, Scheduling space-ground communications for the Air Force Satellite Control Network. J. Sched. **7**(1), 7–34 (2004)
3. F. Marinelli, F. Rossi, S. Nocella, S. Smriglio, A Lagrangian heuristic for satellite range scheduling with resource constraints. Comput. Oper. Res. **38**(11), 1572–1583 (2005)
4. M. Schmidt, Ground station networks for efficient operation of distributed small satellite systems. PhD Thesis, University of Wurzburg (2011)

5. L.V. Barbulescu, Oversubscribed scheduling problems. Thesis Proposal (2002)
6. H. Jung, M. Tambe, A. Barret, B. Clement, Enabling efficient conflict resolution in multiple spacecraft missions via DCSP, in *Proceedings of the 3rd NASA Workshop on Planning and Scheduling* (NASA, Houston, 2002)
7. S.E. Burrowbridge, Optimal allocation of satellite network resources. MSc Thesis, Virginia Polytechnic and State University (1999)
8. M.Y. Kovalyov, C.T. Ng, T.C. Edwin, Fixed interval scheduling: models, applications, computational complexity and algorithms. Eur. J. Oper. Res. **178**, 331–342 (2007)
9. A.J. Vazquez, R.S. Erwin, On the tractability of satellite range scheduling. Optim. Lett. **9**(2), 311–327, Springer (2015)
10. D.A. Vallado, *Fundamentals of Astrodynamics and Applications.* Space Technology Library (Microcosm Press, Hawthorne, CA, 2001)
11. D.A. Parish, A genetic algorithm approach to automating satellite range scheduling. Master Thesis, Air Force Institute of Technology (1994)
12. R.L. Graham, E.L. Lawler, J.K. Lenstra, A.H.G. Rinnoy, Optimization and approximation in deterministic sequencing and scheduling: a survey. Ann. Discret. Math. Discret. Optim. II **5**, 287–326 (1979)
13. M.R. Garey, D.S. Johnson, *Computers and Intractability: A Guide to the Theory of NP-Completeness* (Freeman, New York, 1979). ISBN-13 978-0-7167-1044-8
14. E.M. Arkin, E.B. Silverberg, Scheduling jobs with fixed start and end times. Discret. Appl. Math. **18**, 1–8 (1987)
15. M.R. Fellows, D. Hermelin, F. Rosamond, S. Vialette, On the parametrized complexity of multiple-interval graph problems. Theor. Comput. Sci. **410**(1), 53–61 (2009)
16. A.W.J. Kolen, J.K. Lenstra, C.H. Papadimitrou, F.C.R. Spieksma, Interval scheduling: a survey. Nav. Res. Logist. **54**(5), 530–543 (2007)

Chapter 4
Optimal Satellite Range Scheduling

In Chap. 3 we have presented several variants of the SRS problem, and we have shown that the discretized version of the general problem can be solved in polynomial time for a fixed number of scheduling entities.

In this chapter we will present an algorithm providing the optimal solution to the discretized version of the problem $\mathfrak{R} \mid \mathfrak{r}_j, \mathfrak{p}_{ij} \leqslant \mathfrak{p}_{ij} \leqslant \overline{\mathfrak{p}_{ij}}, C_{\Sigma} \mid \sum \mathfrak{w}_j \mathfrak{U}_j$ (§3.3), which is the most general case, with multiple entities, discrete time, slack, no-precedence, and priorities. We will first provide an algorithm for the subproblem $\mathfrak{R} \mid \mathfrak{r}_j, \overline{\overline{\mathfrak{p}_{ij}}}, C_{\Sigma} \mid \sum \mathfrak{w}_j \mathfrak{U}_j$ (§4.2), and then we will extend this algorithm to cover the variable slack case (§4.3.1) and the redundancy case (§4.3.2).

These algorithms provide an optimal solution in tractable time for a fixed number of entities, and they are based on one from general scheduling from [1], which has been modified to fit SRS constraints, and improved for a faster execution.[1]

4.1 Scenario Model for Fixed Interval SRS

In this section we summarize the model for the fixed interval (or no-slack) problem $\mathfrak{R} \mid \mathfrak{r}_j, \overline{\overline{\mathfrak{p}_{ij}}}, C_{\Sigma} \mid \sum \mathfrak{w}_j \mathfrak{U}_j$.

Let $S = \{s_h\}$ be a set of satellites, $G = \{g_i\}$ a set of ground stations, t_0 a time instant, and T a time window such that $t \in [t_0, t_0 + T]$. We consider the general case regarding the number of resources, so that both $|S| \geqslant 1$ and $|G| \geqslant 1$ (*multiple resource*). We also consider *no-preemption* and *no-precedence* (§3.1.3, Definition 6).

---

[1]This chapter is strongly based on our work [2], which has been extended, reorganized, and integrated with the rest of the book. For a detailed list of the new contributions please see §1.5.

© Springer International Publishing Switzerland 2015
A.J. Vázquez Álvarez, R.S. Erwin, *An Introduction to Optimal Satellite Range Scheduling*, Springer Optimization and Its Applications 106,
DOI 10.1007/978-3-319-25409-8_4

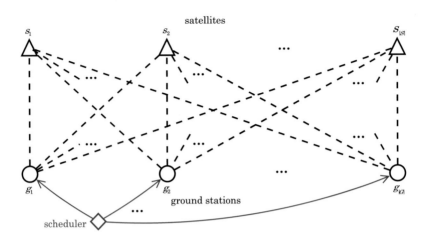

Fig. 4.1 Centralized scheduler

Let $P = \{p_l\}$ be a set of $|P| = N$ passes, with $p_l = (s_h, g_i, t_{s_l}, t_{e_l}, w_l) : w_l \in \mathbb{R} \cap [0, 1]$, so that the problem has *no-slack* and *priorities* (§3.1.3, Definitions 3 and 8). We consider the *no-redundancy* case, so the solution schedule cannot have conflicts among passes (§3.1.3, Definition 5).

As introduced previously, the scheduler is centralized, so that we assume that all the entities will deterministically follow the same schedule, as shown in Fig. 4.1. The set of passes P represents the times where the pairs ground station–satellite can communicate.

Let the set of passes P be the input of the problem, the objective of the fixed interval SRS problem is finding an optimal schedule P^*, which is a feasible schedule with maximal metric (§3.1.4, Definition 10).

4.2 Optimal Solution for Fixed Interval SRS

In Chap. 3 (§3.2 and §3.3) we showed the relations between general scheduling and SRS. Based on these results, we now present an algorithm strongly based on the one in [1] (Theorem 3) for general scheduling. The referenced algorithm generates a graph representing all the feasible states (represented by nodes modeling the state of all the machines) and transitions (represented by edges among nodes) of the system. The referenced algorithm cannot be however directly applied in the SRS problem since it would not take into account conflicts on one satellite being tracked by two different ground stations (C_Σ), but we will introduce the necessary modifications to solve the SRS problem. The main differences are in:

(i) The modifications in the extension of the nodes with new events (4.2.5) for avoiding the existence of unfeasible states in the diagram.

(ii) The propagation of the weights of the passes during the graph creation [(4.2.5) and (4.2.7)].

(iii) The dynamic calculation of the longest path during the graph creation through the avoidance of certain edges (4.2.7).

Note that modification *(i)* avoids both conflicts in ground stations and satellites, as in the referenced algorithm only conflicts of a kind are detected. The performance of the algorithm has been improved with modifications *(ii)* and *(iii)*, and it is also benefited from the fact that passes are associated to only one ground station and one satellite.

4.2.1 Description of the Algorithm

The algorithm can be broken down into three main steps:

1. *Event generation*, where the initial set of passes is converted into a set of events. These events define the stages for the generated graph.
2. *Graph creation*, where nodes and edges are progressively created following the list of events, so that two kinds of iterations exist depending on whether the event is associated to an *start time* or to an *end time* of a pass.
3. *Longest path calculation*, which provides the set of nodes with maximal metric, and thus the optimal schedule. As it will be detailed later, the way the graph is created allows for a dynamic calculation of the path.

In the following subsections we describe formally this process. For a detailed example of the algorithm see §4.5.

4.2.1.1 Event Generation

Similarly to the algorithm in [1], each pass in P is mapped into a pair of events $e = \{t, \phi, s, g, w_x\}$, defined by their start and end times ($t \in \{t_s(p_l), t_e(p_l)\}$), a sign ($\phi \in \{+1, -1\}$ regarding if it is a start or an end time), a ground station $g \in \{1, 2, \ldots, k_2\}$, a satellite $s \in \{0, 1, 2, \ldots, k_1\}$ (the element 0 applies when the ground station is idle), and a priority w_x.

Without loss of generality and to keep the notation compact, ground stations are assumed to be the scheduling resources (or machines in general scheduling problems), although they can be interchanged.

The two bijective functions $f_e^+(p_l) = e^+$ and $f_e^-(p_l)) = e^-$ are defined for separating between start time e^+ and end time e^- events:

$$e^+ = (t_s(p_l), +1, s(p_l), g(p_l), w(p_l)), \tag{4.2.1}$$

$$e^- = (t_e(p_l), -1, s(p_l), g(p_l), 0). \tag{4.2.2}$$

Applying the functions f_e^+ and f_e^- to a set of passes P yields the two sets of events E^+ and E^-, respectively. Let E be the set of $2N$ events generated from P, sorted by ascending time.

$$E = \langle \{e_i\} \rangle \; : \; e_{i-1} \prec e_i \Leftrightarrow t(e_{i-1}) < t(e_i), \tag{4.2.3}$$
$$e_i \in E^+ \cup E^-.$$

4.2.1.2 Graph Creation

The graph generation will be performed in stages (or layers, as in [1]). Every event e_i will be associated to an stage Z_i, and these stages can be seen as sets of nodes. The nodes represent the status of all the ground stations (scheduling entities), so each node n_j will be a vector with their associated satellites.

$$n_j = (s(n_j, g_1), s(n_j, g_2), \dots, s(n_j, g_{k_2})) \; : \tag{4.2.4}$$
$$s(n_j, g_i) \in \{0, 1, 2, \dots, k_1\},$$

where $s(n_j, g_i)$ is the satellite assigned to the ith ground station in node n_j.

Let the *frontier* B_{i-1} be the set of nodes that are checked for modification or deletion at any stage Z_i during the graph generation.

In the *first stage* Z_0 there is only one node, wherein all the resources are empty $\exists^* n_0 \in Z_0; |Z_0| = 1; n_0 = (0, 0, \dots, 0)$, the frontier includes only this node $B_0 = \{n_0\}$, and for the consistency of the algorithm the previous frontier is empty $B_{-1} = \emptyset$ and the edge from n_0 to $\emptyset$ has null weight $\exists^* v = (n_0, \emptyset, 0)$.

For the stage Z_i, nodes n and edges v are generated based on the nodes of the frontier B_{i-1} associated to the previous stage Z_{i-1}, and on the event e_i associated to the current one Z_i. We will use the symbol $\triangleq$ for new assignments.

4.2.1.2.1 Start Time Event Stages

Nodes are generated from those in the frontier wherein the ground station g indicated in the event is idle, so that new nodes keep their state unmodified but the resource $g(e_i)$, which takes its value $(s(e_i))$ from the event.

$$\forall n_j \in B_{i-1} : s(n_j, g(e_i)) = 0,$$
$$s(n_j, g') \neq s(e_i) \; \forall g',$$
$$\exists^* v' = (n_j, n_x, w_x), \; n_x \in B_{i-2},$$

if $\phi(e_i) > 0$, then: $\tag{4.2.5}$

$$\exists^* n_l \in Z_i : s(n_l, g(e_i)) \triangleq s(e_i),$$
$$s(n_l, g') \triangleq s(n_j, g') \; \forall g' : g' \neq g(e_i),$$

$$\exists^* v \triangleq (n_l, n_j, w_x + w(e_i)).$$

The frontier of the new stage includes all the nodes of the frontier from the previous stage plus all the nodes in the new stage.

$$B_i \triangleq B_{i-1} \cup Z_i. \tag{4.2.6}$$

In plain words, for start time events, all the nodes in the frontier are evaluated: if the evaluated node has the ground station indicated in the event idle, and the satellite indicated in the event is not assigned to any ground station, then we can create a new node with the same status as the previous except for the new assignment between the entities in the event. This new node will be connected to the previous by an edge which priority is the sum of the previous one plus the priority of the event. The new frontier will be created through the extension of the previous one with the nodes created in this stage.

Note the new condition on $s(n_j, g')$ in (4.2.5) contrasting to the referenced algorithm, as this last one does not consider this kind of conflicts among passes (or *jobs* in general scheduling literature). Note also the introduction of the propagation of the weights $w_x + w(e_i)$ in the creation of the edges v.

4.2.1.2.2 End Time Event Stages

In these stages, one node is created for each pair of nodes with all the resources keeping the same state except the one indicated in the event, and one edge is created from the one of the pair that has the highest accumulated weight in its path to this new node.

$$\forall n_j \in B_{i-1} : s(n_j, g(e_i)) = s(e_i),$$
$$\exists^* v' = (n_j, n_x, w_j), \ n_x \in B_{i-2},$$

and if $\phi(e_i) < 0$, then also:

$$\exists^* n_y \in B_{i-1} : s(n_y, g(e_i)) = 0,$$
$$s(n_y, g') = s(n_j, g'), \ \forall g' : g' \neq g(e_i),$$
$$\exists^* v'' = (n_y, n_z, w_y), \ n_z \in B_{i-2}, \tag{4.2.7}$$

hence:

$$\exists^* n_l \in Z_i : n_l \triangleq n_y,$$

$$\exists^* v \triangleq \begin{cases} (n_l, n_j, w_j), & \text{if } w_j \geq w_y, \\ (n_l, n_y, w_y), & \text{if } w_j < w_y. \end{cases}$$

At the end of the stage the new nodes are added to the frontier, and the evaluated ones deleted. Let $A_i(l) \triangleq \{n_l\}$ and $D_i(l) \triangleq \{n_j, n_y\}$ be the added and deleted sets from each new node n_l at stage Z_i, then the frontier B_i can be expressed as follows:

$$B_i \triangleq \left\{ B_{i-1} \cup \bigcup_l A_i(l) \right\} - \bigcup_l D_i(l). \tag{4.2.8}$$

That is, for end time events, all the nodes in the frontier are evaluated: if the evaluated node has the satellite indicated in the event assigned to the ground station indicated in the event, then there will also be another node sharing the same status except for the ground station indicated in the event which will be idle. This last node will be duplicated to the new frontier, and an edge from this new node to the one with highest weight will be created. Therefore the new frontier will be created extending the previous one with these new nodes, and all the pairs of selected nodes will be deleted.

Note the selection of the edge with the highest weight w_j or w_y in (4.2.7), which dismisses suboptimal paths, contrasting to the referenced algorithm which postpones the calculation of the longest path after generating the whole graph. In this sense note also the propagation of this highest weight to the new edge v.

4.2.1.3   Longest Path Calculation

All the edges are directed from one stage to a previous one, no node is duplicated at start time event stages, and only one edge of the two that would merge is created at end time event stages; so that the generated graph is a tree with edges connecting nodes from the leaves towards the root (which is the start node n_0).

Backtracking a path from any of the leaves in the tree is trivial due to the unitary outdegree of all the nodes in the graph:

$$\deg^+(n_j) = 1 \quad \forall j \geqslant 0. \tag{4.2.9}$$

Thus, the longest path is obtained in the backtracking process starting at the last created node.

4.2.2   Optimality of the Solution and Complexity of the Algorithm

In this section we provide the proof for optimality of the solution and show the computational complexity of the algorithm.

Theorem 4.1. *The SRS problem with fixed number of ground stations or satellites and a set of passes $P = \{p_1, p_2, \ldots, p_N\}$ with associated weights w_i and fixed times $(t_{s_i}, t_{e_i}) \ \forall p_i \in P$ can be solved in $O(N(k_1 + 1)^{k_2})$ by the algorithm in §4.2.1, where k_1 is the number of satellites or ground stations and (the fixed number) k_2 is the complementary, and N is the number of passes.*

Proof. We will prove optimality of the solution by revision of the properties of the generated graph, and polynomial time solvability by accounting for all the necessary steps of the algorithm for the worst case.

Suppose we relax the node creation in (4.2.7) to create both edges (to n_y and n_j). By definition, according to the algorithm description (§4.2.1.2), adding more nodes or edges to the graph would bring unfeasible states or transitions, and deleting nodes or edges would remove feasible states or transitions. Now, if we delete this relaxation, we will delete unoptimal sub-paths but keep the optimal one.

Since this graph is directed (edges are oriented) and acyclic ($\exists v = (n_x, n_y, w) \Rightarrow n_x \in Z_i,\ n_y \in Z_j,\ i > j$), and there are both a start and an end nodes, a longest path algorithm can be applied to obtain the optimal set of states in polynomial time [1].

This set of nodes in the longest path can be easily transformed into the set of associated passes. By definition (§4.2.1.1), there is a bijection between the passes in P and the events in E^+. Let Z^+ be the subset of all the stages Z associated to events in E^+ (also related by a bijective association). Given that $\exists v = (n_x, n_y, w) \Rightarrow n_x \in Z_i$, $n_y \in Z_j,\ i > j$ and also $n_j \in Z_x, Z_y \Leftrightarrow Z_x = Z_y$, there is a biunivocal association between the subset of nodes in Z^+ in any path and the subset of passes that generated the associated stages. Therefore this is applicable to the path between the end and the start nodes, completing the proof of optimality.

The worst case for the number of nodes in a frontier occurs when all the satellites $k_1 = |S|$ can be assigned to all the ground stations $k_2 = |G|$. Considering also the idle state, an upper bound for the number of nodes is $\max(|B_i|) = (k_1 + 1)^{k_2}$ (§4.2.1.1). Since there are $2N$ stages (actually $2N + 1$, but first and last nodes can be grouped together inside an only "worst case frontier"), an upper bound to the number of states checked during the graph creation is:

$$\max \left\{ \sum_{i=1}^{2N} |B_i| \right\} < 2N(k_1 + 1)^{k_2}. \tag{4.2.10}$$

Even if the ordering of the $2N$ events is taken into account, existing algorithms can provide an ordered set E in $O(2N \log(2N))$ [3]. Even the easiest non-trivial problem $k_1 = 2,\ k_2 = 1$ allows for a relatively high number of events holding the inequality $\log(2N) < (k_1 + 1)^{k_2}$, so that we can dismiss the complexity of the ordering for practical problems.

Furthermore, since all the edges are directed from stage Z_i to stage $Z_j : i > j$, and from (4.2.9) all the nodes have only one edge, backtracking the longest path is trivial starting from the end node (§4.2.1.3). Note that given that the nodes keep the weight of the path to the initial node, the latest generated node has the highest value. Since the maximum number of nodes to backtrack is $2N$ (as many as stages), the process can be done in $O(N)$. Thus the algorithm runs in $O(N(k_1 + 1)^{k_2})$.   □

It is possible that the generated directed acyclic graph (DAG) has frontiers with only one node, wherein all the resources are idle. In this case the DAG can be split into several subgraphs (separated by these frontiers), which can be solved independently. This allows to provide results earlier by serialization or parallelization.

An example on the generation of the graph is provided in §4.5, and simulation results of an implementation of the algorithm are provided in §4.6.

4.3   Extension of the Algorithm

In this section we extend the presented algorithm to solve the variants of the problem
with slack and with redundancy.

4.3.1   Optimal Discretized Variable Slack SRS

According to the notation followed in Chap. 3 (§3.3), the presented algorithm (§4.2)
solves the problem $\mathfrak{R} \mid \mathfrak{r}_j, \overline{\overline{\mathfrak{p}_{ij}}}, C_\Sigma \mid \sum \mathfrak{w}_j \mathfrak{U}_j$. The algorithm can be extended to solve
the discretized version of $\mathfrak{R} \mid \mathfrak{r}_j, \mathfrak{p}_{ij} \leqslant \mathfrak{p}_{ij} \leqslant \overline{\mathfrak{p}_{ij}}, C_\Sigma \mid \sum \mathfrak{w}_j \mathfrak{U}_j$ for a fixed number of
satellites and ground stations.

Note that in the slack problem, given that we are considering no-preemption,
passes generated from the same request will be conflicting even if they are not time
overlapping. This will translate in a growth in complexity of the encoding of the
states of the system, as we show in Corollary 4.1.

Corollary 4.1. *The discretized SRS problem with fixed number of ground stations
and satellites and a set of requests $J = \{j_j\}$ with priorities and no-preemption can
be solved in $O(NM(M+1)^{k_1 k_2})$ by an extension of the algorithm in §4.2.1, where N
is the number of requests, M is the square of the ratio between the longest duration
of the requests and the discretization step, k_1 is the fixed number of satellites and k_2
is the fixed number of ground stations.*

Proof. The first step is to apply Transformation 1 (§3.1.3) to obtain a set of passes P
from the initial set of requests J, that is $P = D_n(J)$. Let $N = |J|$, and let the number
of passes generated from each request be $M = \lfloor \Delta t^{-1} \overline{\overline{\rho}} \rfloor^2$ from Proposition 3.1, so
that the total number of passes is $|P| = NM$.

As in Chap. 3 (§3.1.3), let $D_n^{-1}(p_k)$ be the request originating the pass p_k (that is
$D_n^{-1} : p_k \longrightarrow j_j : p_k \in D_n(j_j)$). From (3.1.17)–(3.1.27) a feasible schedule cannot
contain two passes p_i, p_j such that $D_n^{-1}(p_i) = D_n^{-1}(p_j)$. In order to avoid including
two passes from the same request, we will extend the encoding of the nodes (4.2.4)
to keep track of active requests. We consider that an *active request* is that which
has an associated pass in the schedule associated to a given node, if this node is
associated to an stage earlier than the due time of the request associated to that pass.
Let $P(n_j)$ be the set of passes resulting from backtracking from the node n_j to n_0 and
$J(n_j)$ the set of requests associated to these passes. We redefine the nodes:

$$n_j = \{(s(n_j, g_1), s(n_j, g_2), \dots, s(n_j, g_{k_2})), P_j^a, J_j^a\} :$$
$$s(n_j, g_i) \in \{0, 1, 2, \dots, k_1\}, \tag{4.3.1}$$

where $s(n_j, g_i)$ is the satellite assigned to the *i*th ground station in node n_j, and P_j^a
(J_j^a) is the subset of $P(n_j)$ ($J(n_j)$) associated to requests with due times later than the
time associated to the node n_j. For the initialization $P_0^a = \emptyset$ and $J_0^a = \emptyset$.

Let $j(e_i)$ and $p(e_i)$ be the request and the pass associated to event e_i. We modify the creation of *start time stages* (4.2.5) for avoiding conflicts for same request passes (we include the check $j(e_i) \notin J_j^a$) and for including the active satellite if the node is created (we extend the sets of passes and requests $P_l^a = P_j^a \cup p(e_i)$, $J_l^a = J_j^a \cup j(e_i)$).

We also modify *end time stages* for finding those nodes where the pass associated to the event is active (we add to the check only nodes where the pass associated to the event $p(e_i) \in P_j^a$ is tracked). Those nodes will be transformed, so that new versions will be included in the frontier and the old ones deleted. The new nodes will have the same state as the old ones except for $s(n_l, g(e_i))$, which is now idle, and for the set of active passes, $P_l^a = P_j^a - p(e_i)$ (the set of active requests J_j^a remains unchanged). For the new frontier the successfully evaluated nodes are deleted and the new ones included as in (4.2.8).

Finally, for those end time events associated to the last pass p_k of each request, we have to remove the associated requests $D_n^{-1}(p_k)$ from the sets J_j^a associated to all the nodes in the frontier, and merge nodes with the same state as for the rest of end time stages.

Given that we are representing all the feasible states of the system (taking into account the required checks for no-preemption) and deleting suboptimal transitions, the schedule associated to the set of passes in the longest path is optimal, as in Theorem 4.1. The maximum number of states per frontier in this case needs to take into account the recorded active requests. At the same time there may be at most $k_1 k_2$ time overlapping requests. Each request has associated M passes, and given that passes from the same request are conflicting, only one of them (or none, so that we take $M + 1$) may be on a feasible schedule. Then, taken into account the case where no pass from the request is included in the schedule, for each frontier there will be at most $(M + 1)^{k_1 k_2}$ possible feasible sets P_j^a. Given that there are $2NM$ stages, the algorithm runs in $O(NM(M + 1)^{k_1 k_2})$. $\qquad\square$

4.3.2 Optimal Fixed Interval SRS with Redundancy

The presented algorithm (§4.2) can also be extended to solve the no-slack problem with redundancy, that is $\mathfrak{R} \mid r_j, \overline{\overline{p_{ij}}}, C_x \mid \sum w_j \mathfrak{U}_j$ for a fixed number of satellites and ground stations, and where these entities have m-ary capacities.

Corollary 4.2. *The SRS problem with fixed number of ground stations and satellites and a set of passes $P = \{p_1, p_2, \ldots, p_N\}$ with associated weights w_i and fixed times $(t_{s_i}, t_{e_i}) \; \forall p_i \in P$, and where ground stations (or satellites) have m-ary capacity can be solved in $O(N(k_1 + 1)^{k_1 k_2 + 1})$ by an extension of the algorithm in §4.2.1, where k_1 is the number of ground stations or satellites (which have m-ary capacity) and k_2 is the complementary (which have unitary capacity), and N is the number of passes.*

Proof. The idea of the proof is to model each m-capacity entity as m equivalent independent entities. Let k_1 be the number of satellites, k_2 the number of ground

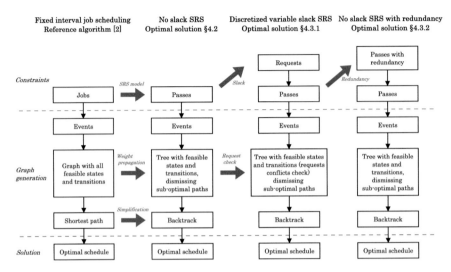

Fig. 4.2 Relations between the algorithms for optimal SRS

stations, and m the maximum number of satellites to which a ground station can communicate at the same time. We can model m-capacity ground stations as m fictitious ground stations ($g_i \rightarrow \{g_{i,1}, g_{i,2}, \ldots, g_{i,m}\}$), so that we create a new set of passes from the initial set P, where each pass is transformed into m passes as follows:

$$\forall p_l \in P : p_l = (s_h, g_i, t_{s_l}, t_{e_l}, w_l) \Rightarrow \exists \{p_{l,1}, p_{l,2}, \ldots, p_{l,m}\}, \qquad (4.3.2)$$

where $p_{l,j} = (s_h, g_{i,j}, t_{s_l}, t_{e_l}, w_l)$. Let P' be the new set of passes, and G' the new set of ground stations. It is easy to see that this transformation is polynomial, and that $|P'| = m|P|$ and $|G'| = m|G|$.

Passes are considered conflicting if they are time overlapping for the same ground station or satellite (passes associated to $g_{i,j}$ and $g_{i,k}$ with $j \neq k$ are not conflicting), as in §3.1.3. The set P' corresponds to a scenario with no-redundancy constraints, and therefore the optimal solution can be obtained by applying the algorithm in §4.2. The algorithm would run in $O(Nm(k_1 + 1)^{mk_2})$, or $O(Nk_1(k_1 + 1)^{k_1k_2})$ for the worst case, which we can bound with $O(N(k_1 + 1)^{k_1k_2+1})$ for shorter notation. □

Note however that if the ground stations had unlimited capacity, that is $m = k_1$, the problem would reduce to solving k_2 independent single entity SRS problems.

Figure 4.2 represents the relations between the referenced algorithm [1] for solving $\Re \mid r_j, \overline{p_{ij}} \mid \sum w_j \mathfrak{U}_j$, the algorithm providing the optimal solution for the fixed interval problem $\Re \mid r_j, \overline{p_{ij}}, C_\Sigma \mid \sum w_j \mathfrak{U}_j$ (§4.2), the extended algorithm for solving the discretized variable slack problem $\Re \mid r_j, p_{ij} \leqslant p_{ij} \leqslant \overline{p_{ij}}, C_\Sigma \mid \sum w_j \mathfrak{U}_j$ (§4.3.1), and the extend algorithm for solving the no-slack problem with redundancy $\Re \mid r_j, \overline{\overline{p_{ij}}}, C_x \mid \sum w_j \mathfrak{U}_j$ (§4.3.2). As explained previously, the algorithm presented in §4.2 incorporates the SRS scenario model, and creates a tree propagating the cumulated weight of the path, dismissing suboptimal paths; the extension presented in §4.3.1 adds the request check for avoiding multiple passes generated from the same request to be in the final schedule; and the extension presented in §4.3.2 removes the redundancy through the creation of fictitious unitary capacity entities.

4.4 Remarks on the Complexity

In this section we present some remarks on the complexity of the algorithms.

4.4.1 Greedy Earliest Deadline Algorithm

As pointed out in the previous chapter (§3.4.1), according to [4] the greedy algorithm provides the optimal solution to the single entity problem with no-slack and no-priorities $(1 \mid r_j, \overline{\overline{p_{ij}}} \mid \sum \mathfrak{U}_j)$. This algorithm selects, given a list of passes ordered by increasing end time, the next pass that can be added to the schedule keeping it feasible, and mark as unfeasible all the passes in the list that are conflicting with it. Its worst case complexity is $O(N^2)$.

However the greedy algorithm does not provide the optimal schedule in general for the multiple entity version of this problem $(\mathfrak{R} \mid r_j, \overline{\overline{p_{ij}}}, C_\Sigma \mid \sum \mathfrak{U}_j)$ nor for the SiRRSP with priorities $(1 \mid r_j, \overline{\overline{p_{ij}}} \mid \sum w_j \mathfrak{U}_j)$. For the first case, a simple example is provided in a scenario with three passes, all mutually conflicting but second with third. In this case the first one would be selected, whereas the optimal decision would have been selecting second and third ones. For the prioritized case, consider a scenario with only two conflicting passes where the second has a higher priority than the first one. In this case the greedy earliest deadline schedule is the first pass, whereas the optimal schedule would be the second one.

4.4.2 Greedy Maximum Priority Algorithm

The selection of the heuristic is crucial for the performance of the greedy algorithm. Besides the end time ordering heuristic introduced in [4], we also considered for the simulations (§4.6) the ordering based on the priorities of the passes (and end times for passes with equal priority). This ordering is used in [5] for ranking jobs as part of a heuristic algorithm.

Note that this heuristic does not provide an optimal schedule in the FI-SiRRSP with priorities: consider a scenario with three passes where first conflicts with second, second with third, and the priority of the second is greater than others' but smaller than other's sum. In this case the greedy schedule would be the second pass, whereas the optimal one would include the first and the third ones.

4.4.3 About the Topology of the Scenario

Although we provided upper bounds for the complexity of the presented algorithm, actual running time depends on the topology of the scenario. Specifically, complexity will be relaxed as the number of conflicting passes diminishes, or equivalently, with the dispersion of the locations and orbits. This can be easily concluded from the way the graph is generated, taking into account all the possible combinations of tracked passes, as in (4.2.10). Note that if there are no conflicts for either satellites or ground stations, the problem reduces to several single entity problems (§3.1.3).

This is not applicable to the studied suboptimal solution algorithms though. The greedy algorithms could improve their performance as the number of conflicts rise, as for every pass selection a group of passes can be dismissed, leading to a shorter execution time.

4.4.4 About the Number of Passes

Based on our experience with numerical examples, we consider that the motion dynamics of the problem limits the number of passes in a day for a pair ground station–satellite to be finite (e.g., low Earth orbit visibility times) (d passes per day). We consider the scheduling horizon to be finite (T days), we assume that the number of passes can be bounded by $N = O(k_1 k_2 dT)$, where k_1 and k_2 are the number of satellites and ground stations, respectively. For more insight see §4.6.3.

4.4.5 About Partial Results

Another consideration should be taken into account in the performance of the algorithm, and that is the availability of partial results earlier than the final schedule. Practical scenarios will generally allow for time periods where all the scheduling entities are idle. This is translated in the proposed algorithm as a boundary of only one element, and the schedule from the first stage to this with unitary size will remain unchanged through the rest of the graph generation, which means that it can be provided as a final (partial) result. Note that all the passes dismissed before this event will not be in the optimal schedule. Let $P(B_i) \subset P$ be the set of passes associated to backtracking from B_i to Z_0, and let p_z be the pass with latest start time that belongs to $P(B_i)$, then $\forall B_i : |B_i| = 1$ we have that $P(B_i) \subset P^*$, and also that:

$$\forall p_j \in P : t_s(p_j) < t_s(p_z), \ p_j \notin P(B_i) \Rightarrow p_j \notin P^*. \qquad (4.4.1)$$

This property (4.4.1) would also allow to segment the execution of the algorithm and improve the execution times through parallel computing.

For greedy algorithms, passes are iteratively added to the schedule until the end of the execution, but neither of the two guarantees that passes dismissed with earlier start time than passes added to the schedule will not be in the final schedule. It is easy to see however that this is valid for the earliest start time heuristic. We will provide further insight about these behaviors in the simulations (§4.6.4).

4.5 Graph Generation Example

This example aims to explain the proposed algorithm in a simple scenario entailing ground stations g_1 and g_2, satellites s_1 and s_2, and four passes (one for each pair station–satellite). For the sake of simplicity we consider the no-slack case. The pass intervals are represented in Fig. 4.3.

Event Generation

The list of the four passes is $P = \{p_1, p_2, p_3, p_4\}$ which we extend into the set of events $\{e_1, e_2, \ldots, e_8\}$:

$$p_1 = (s_1, g_1, t_1, t_4, w_1) \begin{cases} e_1 = (t_1, +1, s_1, g_1, w_1), \\ e_2 = (t_4, -1, s_1, g_1, 0), \end{cases}$$

$$p_2 = (s_2, g_1, t_3, t_6, w_2) \begin{cases} e_3 = (t_3, +1, s_2, g_1, w_2), \\ e_4 = (t_6, -1, s_2, g_1, 0), \end{cases}$$

$$p_3 = (s_1, g_2, t_2, t_7, w_3) \begin{cases} e_5 = (t_2, +1, s_1, g_2, w_3), \\ e_6 = (t_7, -1, s_1, g_2, 0), \end{cases}$$

$$p_4 = (s_2, g_2, t_5, t_8, w_4) \begin{cases} e_7 = (t_5, +1, s_2, g_2, w_4), \\ e_8 = (t_8, -1, s_2, g_2, 0), \end{cases}$$

where $w_1 = 0.6$, $w_2 = 0.6$, $w_3 = 0.8$, and $w_4 = 0.4$.

We obtain the set E by sorting the events by increasing time, so that:

$$E = \{e_1, e_5, e_3, e_2, e_7, e_4, e_6, e_8\}. \tag{4.5.1}$$

Fig. 4.3 Set of passes

Fig. 4.4 Event generation

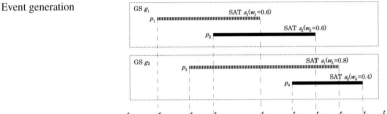

The stages associated to these events are represented in Fig. 4.4. In the following lines we describe the graph generation for the first stages, and then provide the complete graph.

Stage Z_0

For the initialization of the algorithm we have $B_{-1} = \emptyset$, $n_0 = (0, 0)$, $v_0 = (n_0, \emptyset, 0)$, and $Z_0 = B_0 = \{n_0\}$.

Stage Z_1

We examine the first element in E, which is e_1. There is only one node in B_0 to be evaluated, the node n_0, which complies with the conditions in the first paragraph of (4.2.5): (a) the ground station indicated in the event is idle ($s(n_0, g_1) = 0$), (b) the satellite indicated in the event is not assigned to any other ground station ($s(n_0, g') \neq s(e_1) = s_1 \forall g'$), and (c) we have the edge $v_0 = (n_0, \emptyset, 0)$, so that we proceed to create the new node n_1. From the second paragraph of (4.2.5), we duplicate the node n_0, but for the ground station indicated in the event $g(e_1) = g_1$, which had assigned a zero, we now assign $s(e_1) = s_1$, so that $n_1 = (s_1, 0)$. We also create the new edge from the new node (n_1) to the examined one (n_0), with a weight equal to the sum of the edge from n_0 to $\emptyset$ and the weight of the event $w(e_1) = w_1$, so that $v_1 = (n_1, n_0, w_1)$.

There are no more nodes to examine in B_0, so that $Z_1 = \{n_1\}$, and $B_1 = B_0 \cup Z_1 = \{n_0, n_1\}$. The transition from stage Z_0 to Z_1 is displayed in Fig. 4.5.

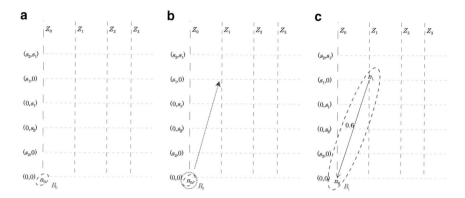

Fig. 4.5 Transition from stage Z_0 to Z_1. (**a**) Initial node. (**b**) Creation of new node. (**c**) Frontier B_1

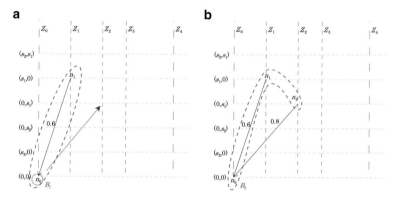

Fig. 4.6 Transition from stage Z_1 to Z_2. (**a**) Creation of new node. (**b**) Frontier B_2

Stage Z_2

The next element in E is the start time event e_5. We evaluate the nodes in B_1, which are n_0 and n_1. The node n_1 cannot be extended, because $s(e_5) = s(n_1, g_1)$, i.e., the satellite indicated in the event is not idle. Note that this condition avoids the conflict wherein two ground stations would be tracking the same satellite. Then we can only extend node n_0, which following the same procedure as in previous stage yields a new node $n_2 = (0, s_1)$, and an edge $v_2 = (n_2, n_0, w_3)$. Then, $Z_2 = \{n_2\}$, and $B_2 = B_1 \cup Z_2 = \{n_0, n_1, n_2\}$. The transition from stage Z_1 to Z_2 is displayed in Fig. 4.6.

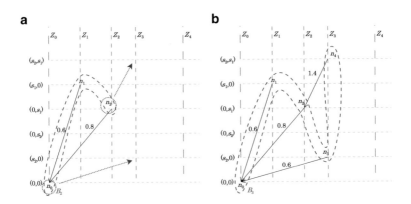

Fig. 4.7 Transition from stage Z_2 to Z_3. (**a**) Creation of new nodes. (**b**) Frontier B_3

Stage Z_3

The next event is also a start time event. We examine n_0, n_1, and n_2 for extension. As for the previous stage, node n_1 is dismissed since $g(e_3) = g_1$ has already assigned s_1, that is, $s(n_1, g(e_3)) = s_1 \neq 0$. Extending nodes n_0 and n_2 we create the pairs $n_3 = (s_2, 0)$, $v_3 = (n_3, n_0, w_2)$ and $n_4 = (s_2, s_1)$, $v_4 = (n_4, n_2, w_3 + w_2)$, so that $Z_3 = \{n_3, n_4\}$ and $B_3 = \{n_0, n_1, n_2, n_3, n_4\}$. The transition from stage Z_2 to Z_3 is displayed in Fig. 4.7.

Stage Z_4

In this case the next event in E corresponds to the end time event e_4. According to the conditions in (4.2.7), we only examine the nodes in B_3 that have the satellite $s(e_4) = s_2$ assigned to the ground station $g(e_4) = g_1$, which is only the node $n_1 = (s_1, 0)$. Note that $v_1 = (n_1, n_0, w_1)$, and also there is an only node in B_3 that has the same state that n_1 except for the only change in the ground station $g(e_4) = g_1$, which is idle, and this node is $n_0 = (0, 0)$, with $v_0 = (n_0, \emptyset, 0)$.

For the pair n_0, n_1, only the node with the highest accumulated weight on its associated vector (v_0 or v_1) will be selected. Since the weight of v_0 is less than that of v_1 we select n_1 as the end of the edge to be created from the new node. We create then the new node $n_5 = (0, 0)$ and the new edge $v_5 = (n_5, n_1, w_1)$. Note that this selection dismisses suboptimal sub-paths.

No more nodes from B_3 can be extended, so the set of created nodes is $\{n_5\}$, and the set of deleted nodes is $\{n_0, n_1\}$. The new frontier is $\{B_3 \cup \{n_5\}\} - \{n_0, n_1\}$, which is $B_4 = \{n_2, n_3, n_4, n_5\}$. The transition from Z_3 to Z_4 is displayed in Fig. 4.8.

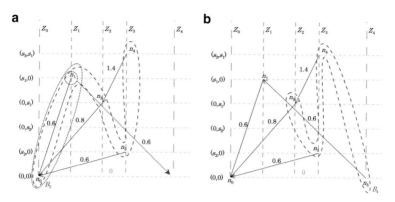

Fig. 4.8 Transition from stage Z_3 to Z_4. (**a**) Creation of new node. (**b**) Frontier B_4

Rest of Stages

Continuing the execution of the algorithm with the rest of the events in E we generate the rest of nodes, edges, stages, and frontiers:

$$
\begin{aligned}
&n_0 = (0,0), &&v_0 = (n_0, \emptyset, 0), &&Z_0 = \{n_0\}, &&B_0 = \{n_0\}, \\
&n_1 = (s_1, 0), &&v_1 = (n_1, n_0, 0.6), &&Z_1 = \{n_1\}, &&B_1 = \{n_0, n_1\}, \\
&n_2 = (0, s_1), &&v_2 = (n_2, n_0, 0.8), &&Z_2 = \{n_2\}, &&B_2 = \{n_0, n_1, n_2\}, \\
&n_3 = (s_2, 0), &&v_3 = (n_3, n_0, 0.6), && && \\
&n_4 = (s_2, s_1), &&v_4 = (n_4, n_2, 1.4), &&Z_3 = \{n_3, n_4\}, &&B_3 = \{n_0, n_1, n_2, n_3, n_4\}, \\
&n_5 = (0,0), &&v_5 = (n_5, n_1, 0.6), &&Z_4 = \{n_5\}, &&B_4 = \{n_2, n_3, n_4, n_5\}, \\
&n_6 = (0, s_2), &&v_6 = (n_6, n_5, 1.0), &&Z_5 = \{n_6\}, &&B_5 = \{n_2, n_3, n_4, n_5, n_6\}, \\
&n_7 = (0,0), &&v_7 = (n_7, n_3, 0.6), && && \\
&n_8 = (0, s_1), &&v_8 = (n_8, n_4, 1.4), &&Z_6 = \{n_7, n_8\}, &&B_6 = \{n_6, n_7, n_8\}, \\
&n_9 = (0,0), &&v_9 = (n_9, n_8, 1.4), &&Z_7 = \{n_9\}, &&B_7 = \{n_6, n_9\}, \\
&n_{10} = (0,0), &&v_{10} = (n_{10}, n_9, 1.4), &&Z_8 = \{n_{10}\}, &&B_8 = \{n_{10}\}.
\end{aligned}
$$

We calculate the longest path walking the graph from the last node: n_{10}, n_9, n_8, n_4, n_2, and n_0. From this set, only nodes n_2 and n_4 correspond to start time events, and specifically to those from passes p_3 and p_2.

We show the generated graph in Fig. 4.9. We mark the longest path in bold lines, and we represent those edges that have not been created (but checked for creation) in light gray dotted lines.

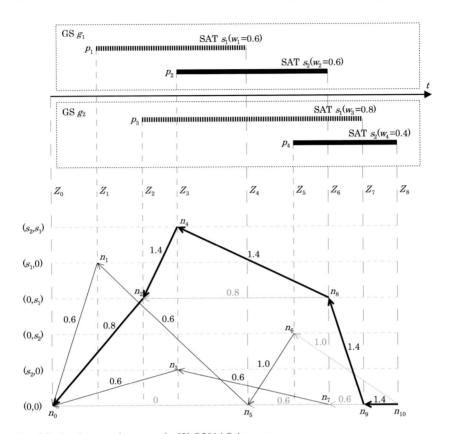

Fig. 4.9 Graph generation example. [2] ©2014 Scitepress

4.6   Simulations

We present two simulations for shedding some light on the influence of the scenario
in the performance of the algorithm. Based on the considerations presented in
§4.4 we compare the algorithm with two greedy heuristics: earliest deadline and
maximum priority. The algorithms have been implemented in MATLAB, and the
simulations were run on a virtual machine with a 3 GHz processor and 2 GB RAM.
For the sake of simplicity we assume that the passes are generated directly from the
LOS times among ground stations and satellites, that is, the no-slack case. That is
generally the case for low Earth orbits, due to the short duration of their associated
visibility windows [6].

   We run the three algorithms on two scenarios, both with $|G| = 5$ stations and
$|S| = 5$ satellites, and varying the size of the scheduling horizon from 1 to 14
days. For the first scenario, which we consider a *practical case*, ground stations
are assigned arbitrary positions and satellites different low Earth orbits; and for the
second scenario, which we consider the *worst case* due to number of conflicts, all the

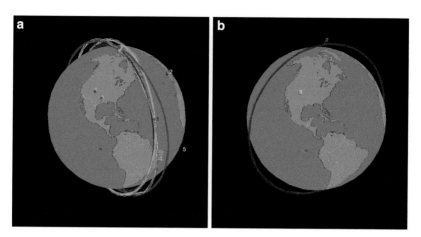

Fig. 4.10 Simulation scenarios. (**a**) Practical case. (**b**) Worst case

locations of the ground stations are identical, and so are all the orbits of the satellites. Furthermore, passes are given a random priority $w_i = v_i/10 : v_i \sim U[1,10]$. We consider a discretization step of 1 min for the calculation of the LOS intervals (start and end times of the passes). The representations of these two scenarios are displayed in Fig. 4.10a, b.

The selected size of the scenarios is enough to show how the dynamics of the problem influence the complexity of the algorithm. As the number of entities increases, shorter execution times are clearly foreseeable for the greedy algorithm compared to the optimal algorithm, based on the complexity bounds of the algorithms.

4.6.1 Simulation: Practical Case

The set of passes and the optimal schedule for the *practical case* scenario are displayed in Fig. 4.11 for a scheduling horizon of 1 day, where passes in the optimal schedule are in blue.

We show in Fig. 4.12a the execution time of the three algorithms as a function of the number of passes in the scheduling horizon with values of $T = \{1, 2, \ldots, 14\}$ days. The graph shows the linear growth (with the number of passes) of the optimal algorithm, and the quadratic growth of the greedy algorithms. Increasing the number of entities would yield higher execution times for the optimal algorithm, however, still outperforming the times of the other algorithms for high $|P|$ values.

In Fig. 4.12b we show the ratio between the studied metric and that of the optimal schedule $\|P^f\|_{\Sigma w} \cdot \|P^*\|_{\Sigma w}^{-1}$. We also show the values of $\|P^*\|_{\Sigma w}$ for all the considered scheduling horizons (*blue*). Note that the priority heuristic achieves near-optimal schedules, but never optimal, for the studied cases.

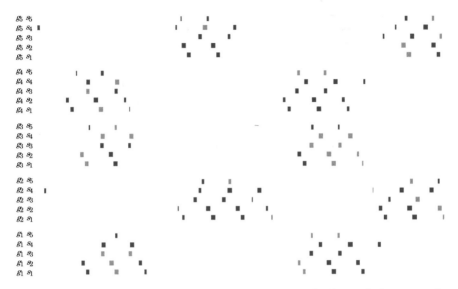

Fig. 4.11 Optimal schedule (*blue*) and dismissed passes (*orange*) for the practical case scenario for a 1 day scheduling horizon

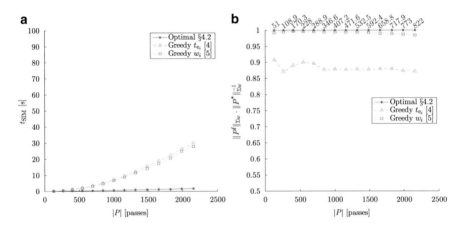

Fig. 4.12 Simulation: practical case. (**a**) Simulation times. (**b**) Metric ratios. [2] ©2014 Scitepress

4.6.2 Simulation: Worst Case

Given that all the locations are identical, and so are all the orbits, this case yields the maximum number of conflicts. Of course this scenario may not correspond to a practical situation, but it will allow to benchmark the worst case behavior of the algorithm. We display the set of passes and the optimal schedule (with its passes in blue) in Fig. 4.13 for a 1 day scheduling horizon.

Fig. 4.13 Optimal schedule (*blue*) and dismissed passes (*orange*) for the worst case scenario for a 1 day scheduling horizon

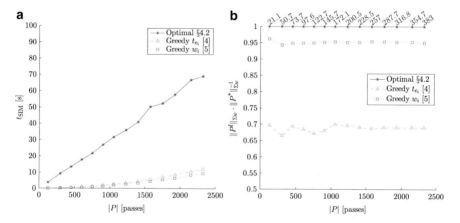

Fig. 4.14 Simulation: worst case. (**a**) Simulation times. (**b**) Metric ratios. [2] ©2014 Scitepress

We show the simulation times in Fig. 4.14a, with similar considerations as those given for the previous scenario. Note that whereas the linear coefficient for the execution time of the optimal algorithm grows compared to the previous scenario, the greedy algorithm reduces its execution time (since for every pass selection a group of passes can be dismissed). However, based on the complexity of these algorithms, as the scheduling horizon extends, the execution time for the greedy-based algorithms will get worse compared to that of the optimal algorithm.

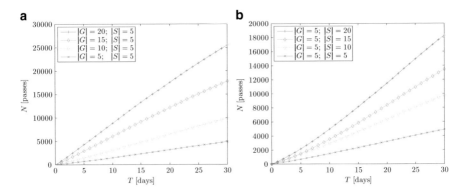

Fig. 4.15 Number of passes and scheduling horizon: extended practical case. (**a**) Varying the number of ground stations. (**b**) Varying the number of satellites

4.6.3 Simulation: Number of Passes

In this section we provide some simulations supporting the remark provided in §4.4.4 on the bound for the number of passes of the scenario. It is out of the scope of this section to provide proof of that remark, however, we consider that these results provide enough justification for its application in this book. We use the same model for the generation of passes and consider the same practical and worst case scenarios used in previous subsections, although we extend their size by adding more entities according to the specifications provided in §4.6 for the two scenarios. We consider scheduling horizons varying from 1 to 30 days, for which we represent the number of passes for different sizes of scenarios. The results for the practical case are displayed in Fig. 4.15, and for the worst case in Fig. 4.16, showing approximately linear growth with T (keeping constant the rest of the parameters) for all cases. Linear growth for variations in $|G|$ or $|S|$ (regular spacing among curves) is only observed in the worst case scenario, due to the repeated orbits and positions.

4.6.4 Simulation: Partial Results

In this section we provide more insight on the behavior of the benchmarked algorithms regarding the availability of partial results. We consider again the practical case and worst case scenarios, but we take a scheduling horizon of 1 day. Figure 4.17a,b show the index, with passes sorted by start time, of those passes iteratively added to the schedule for the practical and worst case, respectively.

According to the previously stated in §4.4.5, the partial schedules provided by the greedy earliest deadline algorithm are not necessarily incremental, but with the considered models for the passes (with similar durations), the time to calculate the

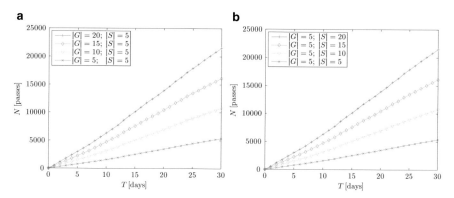

Fig. 4.16 Number of passes and scheduling horizon: extended worst case. (**a**) Varying the number of ground stations. (**b**) Varying the number of satellites

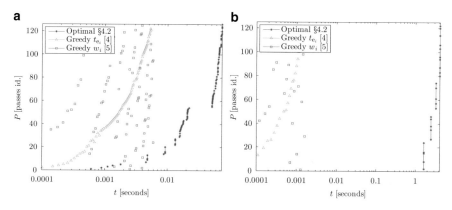

Fig. 4.17 Availability of partial results. (**a**) Practical case. (**b**) Worst case

partial results could be bounded. It can be noted the random order of the passes added to the schedule through the priority heuristic. Also bounds could be provided to this heuristic, but deeper insight on this is out of the scope of this chapter. Finally, the partial schedules provided by the presented algorithm guarantee that passes dismissed before the last pass in the partial schedule will not be in the optimal schedule.

4.7 Summary

We classify the algorithm (§4.2) and its extensions (§4.3) into the set of problems presented in Chap. 3 (§3.5) in Fig. 4.18. Satellite scheduling literature provided an optimal solution to the no-slack and no-priority problem for a single resource

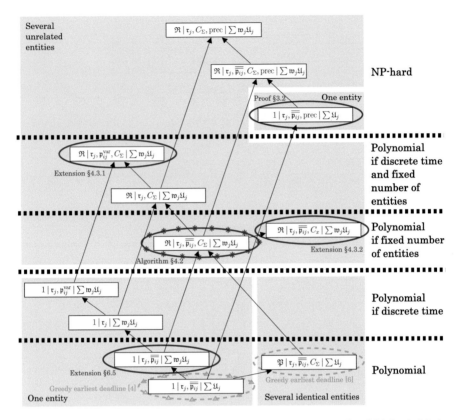

Fig. 4.18 Optimal solutions to the SRS problem, previous literature (*green*) and this book (*blue*)

(greedy algorithm with earliest deadline ordering of the passes [4]) and for multiple identical resources (same algorithm [6]). We have provided the optimal solution to the general case with discretized time and a fixed number of entities.

The presented algorithm for $\mathfrak{R} \mid \mathfrak{r}_j, \overline{\overline{\mathfrak{p}_{ij}}}, C_\Sigma \mid \sum \mathfrak{w}_j \mathfrak{U}_j$ (§4.2) is by definition applicable also to the single scheduling entity problem $1 \mid \mathfrak{r}_j, \overline{\overline{\mathfrak{p}_{ij}}}, C_\Sigma \mid \sum \mathfrak{w}_j \mathfrak{U}_j$. The complexity in this case would be $O(Nk_1)$, given that $k_2 = 1$ and $k_1 \gg 1$. However, an algorithm which solves this problem in $O(N)$ is presented in Chap. 6 (§6.5). Its presentation has been delayed to that chapter since it is based on an approach for a variant of SRS.

Note that the optimal solutions provided in this book are applicable also to their subproblems; and the intractability proofs to the generalizations. Edges in Fig. 4.18. are directed from subproblems to generalizations.

Acknowledgements This research was performed while the author held a National Research Council Research Associateship Award at the Air Force Research Laboratory (AFRL).

References

1. E.M. Arkin, E.B. Silverberg, Scheduling jobs with fixed start and end times. Discret. Appl. Math. **18**, 1–8 (1987)
2. A.J. Vazquez, R.S. Erwin, Optimal fixed interval satellite range scheduling, in *Proceedings of the 3rd International Conference on Operations Research and Enterprise Systems* (Scitepress, Angers, 2014), pp. 401–408
3. D.R. Musser, Introspective sorting and selection algorithms. Softw. Pract. Exp. **27**(8), 983–993 (1997)
4. S.E. Burrowbridge, Optimal allocation of satellite network resources. Master Thesis, Virginia Polytechnic and State University (1999)
5. W.J. Wolfe, S.E. Sorensen, Three scheduling algorithms applied to the Earth observing systems domain. Manag. Sci. **46**(1), 148–168 (2000)
6. L. Barbulescu, J.P. Watson, L.D. Whitley, A.E. Howe, Scheduling space-ground communications for the air force satellite control network. J. Sched. **7**(1), 7–34 (2004)

Part III
Variants of Satellite Range Scheduling

Chapter 5
Noncooperative Satellite Range Scheduling

So far we have formally presented the SRS problem in Chap. 3, and we have provided an algorithm for finding the optimal solution to the problem in Chap. 4. Regardless of the performance of the provided solution, however, satellites may be associated to different operators, and applying a centralized algorithm in such a distributed system could bring situations like a selfish mission operator unilaterally deviating from the initial schedule to improve its own metric (by taking advantage of the knowledge on the expected behavior of the rest of the operators). Without loss of generality operators are removed from the model, so that we consider the satellites as the proxies for their intents and actions.

Game-theoretic approaches have been already applied to general scheduling problems [1–3], but not to SRS problems. In this chapter we model the noncooperative SRS problem from a game-theoretic perspective, in which *players* representing the different missions (satellites) compete for the use of the *resources* in the ground station network (ground stations). We will show that the system converges to a Stackelberg equilibrium solution where the players cannot yield better outcomes unilaterally. Following the slack relations presented in Chap. 3, in this chapter we focus on the fixed interval version of the problem, but these results can be generalized to the variable size window problem through Transformation 1 (§ 3.1.3). At the end of the chapter we introduce limited information versions of this problem.[1]

---

[1]This chapter is strongly based on our work [4], which has been extended, reorganized, and integrated with the rest of the book. For a detailed list of the new contributions please see § 1.5.

© Springer International Publishing Switzerland 2015
A.J. Vázquez Álvarez, R.S. Erwin, *An Introduction to Optimal Satellite Range Scheduling*, Springer Optimization and Its Applications 106, DOI 10.1007/978-3-319-25409-8_5

5.1 Scenario Model for the SRS Game

As in the previous chapter we summarize the basic formulation of the problem. Let $S = \{s_i\}$ be a set of satellites, and $G = \{g_h\}$ a set of ground stations. We consider a finite scheduling horizon T starting at t_0, so that $t \in [t_0, t_0 + T]$.

Let a *pass* p_l be a tuple modeling a visibility time window from a start time t_s to an end time t_e between the satellite s_i and the ground station g_h, and with an assigned priority w_l, that is: $p_l = (s_i, g_h, t_{s_l}, t_{e_l}, w_l) : w_l \in \mathbb{R} \cap [0, 1]$. The term w_l characterizes the weight or *priority* which the satellite s_i associates to that request, normalized between 0 and 1. Let $P = \{p_l\}$ be a set of $|P| = N$ passes.

Whereas the traditional approach would generate a centralized schedule and distribute it to all the ground stations and mission control centers (§ 4.1), here we assume that the ground stations will accept any request from any mission control center if the ground station is idle. Equivalently, there is no possibility for the mission control centers to commit a ground station to track a pass in the future; the only possibility is to request a pass that is about to begin.

The assumption on perfect information for the satellites computing the schedules in a distributed way may seem unrealistic, but note that we are considering them the proxies for the intents and actions of the mission control centers. We will present two variants of the problem with limited information (§ 5.4): uncertain priorities and uncertain passes. Note in this chapter we refer to deterministic uncertainty (reduced information), as opposed to nondeterministic uncertainty treated in the next chapter.

We show in Fig. 5.1a the centralized problem, with a single scheduler for all the entities; in Fig. 5.1b, we show the distributed problem with perfect information, where all the mission control centers are mutually connected; Fig. 5.1c shows the distributed problem with uncertain priorities—that is, each satellite (mission control center) has knowledge of the entire pass list, and their own priorities, but not on the

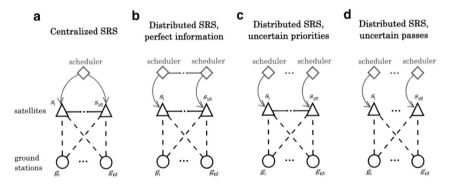

Fig. 5.1 SRS scenarios with different topologies. (**a**) Centralized SRS. (**b**) Distributed SRS with perfect information. (**c**) Distributed SRS with uncertain priorities. (**d**) Distributed SRS with uncertain passes

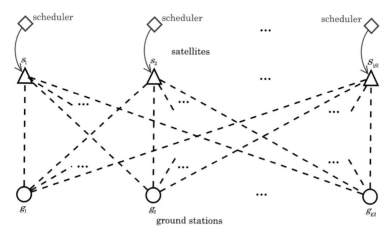

Fig. 5.2 Distributed scheduler

priorities of the other satellites; finally, Fig. 5.1d adds the issue of uncertain passes, where the satellites (mission control centers) only know of those passes involving them directly, and have no knowledge of passes involving the other satellites. Again, note the representation of the information shared among mission control centers as connections between the satellites (for the passes) and their associated schedulers (for the priorities of the passes).

For simplicity, and to keep the parallelism with the topology shown in § 4.1 for the centralized problem, in this chapter we will consider the TVG (time varying graph) model displayed in Fig. 5.2.

5.2 Elements of the SRS Game

In this section we detail the elements of the satellite range scheduling game. The set of passes represents the baseline for the game, wherein the satellites will be the players, taking sequential decisions based on the start times of the passes.

5.2.1 Players

We consider the satellites to be the *players* ($S = \{s_i\}$) in the game. Ground stations are in this case the resources to be scheduled, but their roles could be exchanged.

5.2.2 Sequential Decisions

Players will take *sequential decisions* corresponding to the start times of their associated passes, so that the game will be played from $t = t_0$ to $t = t_0 + T$. Note that the execution time of the schedule could be reduced, so that this game could also be played before the actual scheduling horizon starts. This would only require to scale all the times inside the window T to fit into a smaller window $T_g \ll T$, since only the relative order of the start and end times of the pases is relevant for the presented algorithm.

5.2.3 Actions

The sequential decisions made by the players at the start times of their associated passes will correspond to one of two possible *actions* on these passes: either tracking ($\mathcal{T}$) or dismissing ($\mathcal{D}$) the associated pass. We denote the set of actions for each player as $\mathbf{A}_i = \{\mathcal{T}, \mathcal{D}\} \ \forall s_i$.

5.2.4 Shared Information

We consider three cases in this chapter:

Perfect information The players know the complete set of passes P for the scheduling horizon T, and they also have access to all the actions taken by the other players.

Uncertain priorities In this case the players know the complete set P except for the priorities of the passes associated to other players, and they also have access to the actions taken by the other players.

Uncertain passes This is the worst case information-wise, where the players have no information whatsoever about the other players, only the list of passes associated to themselves.

5.2.5 Payoffs

The *payoffs* can be seen as the fractions of the centralized metric (§ 3.1.4) associated to each player at the terminal nodes, or equivalently, as the sum of the weights of its tracked passes at the end of the game. Let $P' \subset P$ be the executed schedule at the end of the game, and $w(p_l)$ and $s(p_l)$ the priority and the satellite associated to the pass p_l. Then we define the payoff $J_i(P')$ for satellite s_i as follows:

$$J_i(P') = \sum_l w(p_l) \quad \forall p_l \in P' : s(p_l) = s_i. \tag{5.2.1}$$

5.2.6 Rationality

We consider the players to be *rational*, which means that each of them aims at maximizing its associated payoff $J_i(P')$ at the end of the game. Note that for the perfect information case we will assume that all the players know that the rest of the players are rational.

5.2.7 Extensive Form

The *extensive form representation (EFR) of the game* is a graph representing all the actions taken by the players, staggered through levels. These levels are defined by the initial set of passes sorted by increasing start time, so that at each level the satellite associated to that level takes an action: either tracking ($\mathscr{T}$) or dismissing ($\mathscr{D}$) the pass. The executed schedule at the end of the game is required to be feasible. It is easy to see that the $\mathscr{D}$ action will be always feasible, but the $\mathscr{T}$ action will be only feasible if the satellite and the ground station associated to the pass are idle at that time (or equivalently, if the considered pass is not conflicting (§ 3.1.3) with any pass already tracked). At each level, for every vertex and for every feasible action, a new vertex will be created in the next level along with an edge pointing to the previous vertex. Vertexes model the state of the network (which ground station is tracking which satellite), and also the payoffs associated to each satellite, which will incorporate the weight of the passes that are tracked ($\mathscr{T}$) by each satellite.

Let us consider for now the perfect information case. We show that the problem can be formulated as an $|S|$-*person nonzero-sum finite game without chance moves* following the Definition 3.10 in [5]:

1. The extensive form representation of the game has an *initial vertex*, where all the ground stations and all the satellites are idle. This vertex corresponds to the earliest time of the scheduling horizon ($t = t_0$).
2. We assign a vector to each of the vertexes with each element being the sum of the weights of the passes associated to that player (costs are considered in the original definition, but we consider *payoffs* instead without loss of generality). These payoffs can be calculated iteratively from one level to the next one, until the terminal vertexes.
3. The extensive form of the game can be partitioned, being each of these *partitions* or levels associated to the player that corresponds to the pass associated to the next level. That is, we say that a player s_i is associated to level L if the pass associated to the next level $L + 1$ involves the satellite s_i.

4. Given that we assume perfect information, every player knows the moves of
   others, and thus every *information set* is a singleton. Furthermore, we delete
   unfeasible moves without loss of generality to this definition, given that we could
   develop the graph including these unfeasible moves but assign to them arbitrarily
   low payoffs (negative values) for the players on these states. Then these moves
   will never be taken, and thus they can be pruned from the graph.

5.3   SRS Game with Perfect Information

In this section we will present formally the satellite range scheduling game with
perfect information, based on the elements of the game introduced in the previous
section. We will also define its associated Stackelberg equilibrium, and provide the
algorithm for its calculation.

Definition 11. The *SRS game with perfect information* is the extensive form game
$(S, P, A, J, I_{\text{PI}})$ where:

- S is the set of players $\{s_i\}$ (satellites).
- P is the set of passes, which defines the order of play $\langle\{s(p_l)\}\rangle_{t_s}$ (satellites
  associated to the passes sorted by increasing start time) and the partial payoffs
  for each player.
- A is the set of actions available to each player for each pass, which are either
  tracking ($\mathscr{T}$) or dismissing ($\mathscr{D}$) the pass, so that $A = A_i = \{\mathscr{T}, \mathscr{D}\} \ \forall s_i$.
- J is the payoff function, $J_i(P')$ defined for player s_i and schedule P', which is the
  sum of the weights of the passes in that schedule that are associated to s_i.
- I_{PI} is the information available to the players at each level of play, which includes
  both the complete set P, and the history of play H_l (the Cartesian product of the
  actions $a \in A$ for each pass from the start to the current level of play). So for
  each s_i and for each level of play l we have $I_i(l) = \{P, H_l\}$ with $H_l = a_{p_1} \times a_{p_2}$
  $\times \cdots \times a_{p_l} \ \forall a_{p_l} \in A, \forall p_l \in \langle P \rangle_{t_s}$.

Based on the presented model, the centralized SRS problem can be seen as a
subproblem of the distributed SRS problem:

Lemma 5.1. *The noncooperative SRS game with perfect information generalizes
the centralized SRS problem.*

Proof. The centralized problem can be obtained from $(S, P, A, J, I_{\text{PI}})$ if we change
the payoffs of the players (5.2.1) for $J_i(P') = \|P'\|_{\Sigma w} \ \forall s_i \in S$, so that all the players
share the same objective: maximizing the centralized metric of the schedule.   □

Based on the presented model, with the players taking sequential decisions at a
predefined order, the game can be modeled as a Stackelberg game. Assuming that
the players are rational, the leader knows that the follower will always play his best
strategy (depending on the leader's strategy), and the leader will play that strategy
which maximizes her payoff (considering the best response of the follower) [5].

Definition 12. The *Stackelberg equilibrium of the SRS Game with perfect informa-tion* is the subset $P' \subset P$ where no player s_i unilaterally changing its set of actions can reach an alternative schedule $P'' \subset P$ such that $J_i(P'') > J_i(P')$.

The calculation of this equilibrium is non-trivial. There are 2^N possible schedules $P_{sub} \subset P$, which are not necessarily feasible (checking the feasibility of a single schedule takes N^2 for the worst case). And note that simply evaluating these schedules would not be enough, as they are just the terminal nodes of the extensive form of the game, where also the order of play is important for finding this equilibrium.

5.3.1 Description of the Algorithm

In this section we modify the algorithm presented in § 4.2, which provided an optimal solution to the centralized SRS problem in polynomial time. The referenced algorithm generated a graph with nodes representing all the possible states of the system, and edges a subset of all the possible transitions among these states, which contained the optimal solution. The modified algorithm allows to find the Stackelberg equilibrium in the distributed problem with perfect information.

The main difference with the referenced algorithm is on the end time event stages (5.3.11), where rather than on a central metric, decisions on sub-paths are made based on the payoff of the leading player for those sub-paths. We present a summary of the introduced modifications in Fig. 5.3 (§ 5.4.1), and a detailed example on the application of this algorithm in § 5.6.

5.3.1.1 Event Generation

The generation of events is similar as in § 4.2. Passes ($p_l = (s_a, g_b, t_{s_l}, t_{e_l}, w_l)$) are mapped into events $e = \{t, \phi, s, g, w_x\}$, defined by their start and end times ($t \in \{t_s(p_l), t_e(p_l)\}$), a sign ($\phi \in \{+1, -1\}$) regarding if it is a start or an end time, a satellite $s \in \{1, 2, \ldots, |S|\}$, a ground station $g \in \{0, 1, 2, \ldots, |G|\}$ (the element 0 applies when the satellite is idle), and a priority w_l. Note that we change the indexes of s and g from the pass compared to § 4.2.

The two bijective functions $f_e^+(p_l) = e^+$ and $f_e^-(p_l)) = e^-$ are defined for separating between start time e^+ and end time e^- events, and applied to the set of passes P they yield the two sets of events E^+ and E^-, respectively. Let E be the set of $2N$ events generated from P:

$$f_e^+ : p_l \longrightarrow e^+ = (t_s(p_l), +1, s(p_l), g(p_l), w(p_l)), \tag{5.3.1}$$

$$f_e^- : p_l \longrightarrow e^- = (t_e(p_l), -1, s(p_l), g(p_l), 0), \tag{5.3.2}$$

$$E = (\{e_i\}) : e_{i-1} \prec e_i \Leftrightarrow t(e_{i-1}) < t(e_i), \ e_i \in E^+ \cup E^-. \tag{5.3.3}$$

Start time events represent the decisions from the players, and end time events are introduced to reduce the size of the extensive form of the game, as otherwise duplicates of the same node would exist in the same level (nodes representing the same state of the network).

We change the definition of the weight of the events for taking into account the distributed objectives:

$$w(e_i) = (0, \ldots, 0, w(p_l), 0, \ldots, 0), \qquad (5.3.4)$$

where $w(p_l)$ (the weight of the pass p_l) is the $\{s(p_l)\}$th element in the vector (where $s(p_l)$ is the satellite associated to the pass p_l), and the rest of the elements are zero. This is the vector of payoffs of the players, which can be associated to the nodes or to the edges originating on them.

5.3.1.2 Graph Elements

We describe all the different elements of the graph in the following lines.

Stage   The graph generation will be performed in stages. Every event e_i will be associated to an stage Z_i, and these stages can be seen as sets of nodes.

Node   The nodes represent the state of all the satellites (players), so each node n_j will be a vector with their associated ground stations.

$$n_j = (g(n_j, s_1), g(n_j, s_2), \ldots, g(n_j, s_{|S|})), \qquad (5.3.5)$$

with $g(n_j, s_i) \in \{0, 1, 2, \ldots, |G|\}$, and where $g(n_j, s_i)$ is the ground station assigned to the ith satellite in node n_j.

Edge   Edges will be created iteratively among nodes to represent transitions (e.g., a satellite starts to be tracked by a ground station, or a satellite stops being tracked), and each node v_j has the form $v_j = (n_{\text{start}}, n_{\text{end}}, w_j)$, $\forall n_j \; \exists^* v_j$, where n_{start} is the starting node for the edge $(n_{\text{start}} = n_j)$, n_{end} is the ending node, and w_j is the vector of payoffs defined in further paragraphs.

Sub-path We define a subset of passes for keeping track of the sub-schedule (or sub-path in the graph), which will be the set of nodes obtained by backtracking from a current node iteratively to the first node of the graph through the edges:

$$m_j = \{n_0, \ldots, n_j\}, \; \forall n_j \; \exists^* m_j. \qquad (5.3.6)$$

Schedule Let the subset $P(m_j) \subset P$ be the schedule associated to the sub-path m_j, which includes the passes associated only to those nodes belonging to start time stages.

Payoff vector   We define the vector of payoffs w_j as follows:

$$w_j = \left(J_1(P(m_j)),\ J_2(P(m_j)),\ \ldots,\ J_{|S|}(P(m_j))\right). \qquad (5.3.7)$$

The centralized metric of the schedule $\|\cdot\|_{\Sigma w}$ coincides with the sum of all the elements of the payoff vector:

$$\|P(m_j)\|_{\Sigma w} = \sum_{a=1}^{|S|} J_a(P(m_j)). \qquad (5.3.8)$$

Frontier   Let the *frontier* B_{i-1} be the set of nodes that are checked for modification or deletion at any stage Z_i during the graph generation. Note that a frontier could have nodes in different stages.

5.3.1.3 Graph Creation

We show the steps to perform at each stage of the algorithm in the following paragraphs. We use the symbol $\triangleq$ for new assignments.

Initialization   The first step of the algorithm is the initialization. This includes: the first stage Z_0, which has only one node, wherein all the satellites are idle $|Z_0| = 1$, $\exists^* n_0 \in Z_0$, $n_0 \triangleq \overline{0}_{|S|} = (0, 0, \ldots, 0)$, $m_0 \triangleq \emptyset$; and the first frontier $B_0 = Z_0$. For consistency the previous frontier is empty $B_{-1} \triangleq \emptyset$, and the edge from n_0 to $\emptyset$ has null weight $\exists^* v \triangleq (n_0, \emptyset, \overline{0})$.

Stages associated to events   As for the referenced algorithm, we examine the events in E. For the stage Z_i, nodes n and edges v are generated based on the nodes of the frontier B_{i-1} associated to the previous stage Z_{i-1}, and on the event e_i associated to the current one Z_i. Note that there will be a bijection among stages associated to start time events and levels of the game (see § 5.3). The steps to perform at each of these stages depend on whether the stage is associated to a start time event or to an end time event, as we will show in the following lines.

Start Time Event Stages

Nodes are generated from those in the frontier wherein the satellite indicated in the event is idle, so that new nodes keep their state unmodified but the satellite $s(e_i)$, which takes its value $(g(e_i))$ from the event.

$$\forall n_j \in B_{i-1} : g(n_j, s(e_i)) = 0, \ g(n_j, s') \neq g(e_i) \ \forall s',$$
$$\exists^* v' = (n_j, n_x, w_x), \ n_x \in B_{i-2},$$
$$\exists^* m_j,$$

if $\phi(e_i) > 0$, then:

$$\exists^* n_l \in Z_i : g(n_l, s(e_i)) \triangleq g(e_i),$$
$$g(n_l, s') \triangleq g(n_j, s') \ \forall s' \neq s(e_i),$$

$$\exists^* v \triangleq (n_l, n_j, w_x + w(e_i)),$$

$$\exists^* m_l \triangleq m_j \cup n_l.$$

(5.3.9)

The frontier of the new stage includes all the nodes of the frontier from the previous stage plus all the nodes in the new stage.

$$B_i \triangleq B_{i-1} \cup Z_i. \tag{5.3.10}$$

Expression (5.3.9) is similar to its equivalent one in § 4.2, with the only modification on the addition of the history vector m_l (and exchanging the roles of ground stations and satellites).

End Time Event Stages

In these stages, one node is created for each pair of nodes with all the resources keeping the same state except the one indicated in the event. Differently from the referenced algorithm, in this case the two path-vectors m_x and m_y have to be compared to obtain the leader satellite in the path, i.e., that which decides to track a pass earlier, so that the two components of the vectors associated to this entity are compared and the path with the highest payoff for the leader is selected.

$$\forall n_j \in B_{i-1} : g(n_j, s(e_i)) = g(e_i),$$
$$\exists^* v' = (n_j, n_x, w_j), \ n_x \in B_{i-2},$$
$$\exists^* m_j,$$

and if $\phi(e_i) < 0$, then also:

$$\exists^* n_y \in B_{i-1} : g(n_y, s(e_i)) = 0,$$
$$g(n_y, s') = g(n_j, s'), \ \forall s' \neq s(e_i),$$
$$\exists^* v'' = (n_y, n_z, w_y), \ n_z \in B_{i-2},$$
$$\exists^* m_y,$$

(5.3.11)

we define the leader satellite s_{lead} :

$$\Delta P = (P(m_j) - P(m_y)) \cup (P(m_y) - P(m_j)),$$
$$\exists^* p_x \in \Delta P : t_s(p_x) < t_s(p_y) \; \forall p_y \in \Delta P - p_x,$$
$$s_{\text{lead}} = s(p_x),$$

hence:

$$\exists^* n_l \in Z_i : n_l \triangleq n_y,$$

$$\exists^* v_l \triangleq \begin{cases} (n_l, n_j, w_j), & \text{if } w_j(s_{\text{lead}}) \geqslant w_y(s_{\text{lead}}), \\ (n_l, n_y, w_y), & \text{if } w_j(s_{\text{lead}}) < w_y(s_{\text{lead}}), \end{cases}$$

$$\exists^* m_l \triangleq \begin{cases} m_j, & \text{if } w_j(s_{\text{lead}}) \geqslant w_y(s_{\text{lead}}), \\ m_y, & \text{if } w_j(s_{\text{lead}}) < w_y(s_{\text{lead}}). \end{cases}$$

In order to determine the leader satellite s_{lead} first we calculate the difference of the sub-paths associated to the two nodes [ΔP (5.3.11)], from which we take the pass with earliest start time. The satellite associated to that pass is the leader satellite s_{lead}, which is the one that decides on which path to take (that with the highest payoff for her). If the two nodes yield the same payoff to the satellite, it will select the sub-path where the differing pass is actually tracked (m_j).

At the end of the stage the new nodes are added to the frontier, and the evaluated ones deleted. Let $A_i(l) \triangleq \{n_l\}$ and $D_i(l) \triangleq \{n_j, n_y\}$ be the added and deleted sets from each new node n_l at stage Z_i, then the frontier B_i can be expressed as follows:

$$B_i \triangleq \left\{ B_{i-1} \cup \bigcup_l A_i(l) \right\} - \bigcup_l D_i(l). \tag{5.3.12}$$

We emphasize that *it is the leading player (s_{lead}) the one that decides on which sub-path to take, and not simply the satellite associated to the pass of that stage*, as this is the main difference with the algorithm providing the optimal solution (§ 4.2).

5.3.2 Stackelberg Equilibrium Solution

We now provide the proof that the algorithm presented in § 5.3.1 yields a unique Stackelberg equilibrium. According to Definitions 11 and 12, and to the algorithm described in § 5.3.1, it follows that:

Proposition 5.1. *The Stackelberg equilibrium of the SRS game with perfect information is directly the subset of P with the elements indicated in the history vector m' associated to the node n' (that is $m' = (n_0, \ldots, n')$) in the last stage of the graph (that is $\exists^* n' \in Z_{2N}$).*

Proof. We generate the extensive formulation of the game by relaxing the algorithm in § 5.3.1 applying two modifications described below. We will show that the sub-tree generated without these modifications contains the Stackelberg equilibrium:

Modification I: Let us suppose we do not remove unfeasible actions [we delete the condition $g(n_j, s') \neq g(e_i)$ $\forall s'$ from (5.3.9)], to which we now assign arbitrarily low payoffs (in (5.3.9) we change v so that $\exists^* v \triangleq (n_l, n_j, w_x + \omega_a)$ where $\omega_a < -J_a(P)$ where s_a is the satellite associated to the event e_i) to assure that players never select them.

Modification II: Let us also suppose we do not merge sub-paths, so that the last paragraph of (5.3.11) is now:

$$\begin{aligned} \exists^* n_l \in Z_i : n_l \triangleq n_y, \quad \exists^* n_{l+1} \in Z_i : n_{l+1} \triangleq n_y, \\ \exists^* v_l \triangleq (n_l, n_j, w_j), \quad \exists^* v_{l+1} \triangleq (n_{l+1}, n_y, w_y), \qquad (5.3.13) \\ \exists^* m_l \triangleq m_j, \qquad\qquad\quad \exists^* m_{l+1} \triangleq m_y. \end{aligned}$$

Then at the stage where two sub-paths would merge, two potential nodes would be created n_l, n_{l+1} (instead of only n_l), both with the same state ($n_l = n_{l+1}$) but different payoff vectors ($w_l = w_j$, $w_{l+1} = w_y$).

From these two nodes n_l and n_{l+1}, two new trees (ψ_l and ψ_{l+1}) would be generated which correspond to two *strategically equivalent games*, since *(a)* the initial node has the same state ($n_l = n_{l+1}$) and *(b)* the vector of payoffs for every node of one of the two trees is equal to that (in the same position) from the other tree minus the constant vector with the difference between the payoff vectors of the two initial nodes. That is, let $w_{l,l+1} = w_l - w_{l+1}$, then $\forall n_x \in \psi_l$ we have that $w_x = w'_x + w_{l,l+1}$ $\forall n'_x \in \psi_{l+1}$.

Therefore we can delete one of the two trees (as it is done in (5.3.11), so that we are actually undoing *Modification II*), thus reducing the complexity of the problem compared to calculating the equilibrium in the complete tree. We keep the tree starting from the node wherein the payoff for the leader (of the sub-path) [either $w_j(s_{\text{lead}})$ or $w_y(s_{\text{lead}})$ in (5.3.11)] is higher, and in order to obtain this leader we compare the two sub-paths [$P(m_j)$ and $P(m_y)$ in (5.3.11)] associated to the two nodes. This pruning is done sequentially for every end time stage. Due to this independence of the future from the selection of the sub-path by the leading player, these leading players will not deviate from their higher-payoff actions. And given that unfeasible paths have arbitrarily low payoffs, rational players will never select them, so that it is easy to see that we can also undo *Modification I*.

Note that branching for other players' passes between the start time event associated to the leading player and the two potential nodes would only generate more pairs of potential nodes, but the leader would be the same as for the first pair since the difference in the two sub-paths would still be associated to this player.

In order to guarantee the uniqueness of the equilibrium, we assume that in case two nodes have the same payoff for the leader, it will prefer the sub-path where the differing pass is actually tracked [or equivalently m_j is preferred over m_y if $w_j = w_y$ in (5.3.11)].

Thus no player gets a higher payoff by unilaterally deviating from the described algorithm, and all the players arrive to the same equilibrium solution. □

5.3.3 Computational Complexity

In this section we provide a complexity bound for the algorithm.

Proposition 5.2. *The Stackelberg equilibrium for the SRS game with perfect information can be computed in $O(N(\log N)(k_1 + 1)^{k_2})$ by the algorithm in § 5.3.1, where k_1 is the number of resources and (the fixed number) k_2 is the number of players, and N is the number of passes.*

Proof. The algorithm is similar to that from § 4.2, (which is computable in $O(N(k_1 + 1)^{k_2})$, product of the number of stages and the number of nodes in a stage for the worst case) with the only added computation being the calculation of the first differing element in the history vectors [pass p_x in (5.3.11)], which can be done through binary search in $O(\log N)$.

Then for the worst case a similar proof to that from § 4.2.2 can be followed to show that the algorithm runs in $O(N(\log N)(k_1 + 1)^{k_2})$. □

Introducing the considerations about the number of passes introduced in § 4.4.4 in the proof above, we have that for a fixed number of entities the algorithm is polynomial in the size of the scheduling horizon.

Note however that keeping history vectors for every node brings a high space complexity for the algorithm. We can instead backtrack the two sub-paths and then check the first different pass. This increases the complexity of the algorithm from $O(N(\log N)(k_1 + 1)^{k_2})$ to $O(N^2(k_1 + 1)^{k_2})$, but it may be more convenient for numerical simulation. We will use this new approach in the results we present in § 5.7.

5.4 Limited Information Versions of the Problem

In this section we relax the conditions of perfect information to consider two cases of limited information:

Uncertain priorities   We consider in this case that the players, even though they have access to the history of play, they only know the priorities of the passes associated to themselves, and thus the behavior of the rest of the players cannot be predicted.

Uncertain passes   This is the worst case regarding information for the players, which only know the subset of passes associated to themselves. In this case the players will have the uncertainty, before tracking a pass, that this pass may be conflicting with another one already being tracked by another player.

5.4.1 SRS Game with Uncertain Priorities

We provide the formal definition for the SRS game with uncertain priorities.

Definition 13. The *SRS game with uncertain priorities* is the extensive form game (S, P, A, J, I_{UW}) where:

- S is the set of players $\{s_i\}$ (satellites).
- P is the set of passes.
- P_{nw} is the set of passes without their associated priorities, which defines the order of play $\langle\{s(p_l)\}\rangle_{t_s}$ (satellites associated to the passes sorted by increasing start time) and the partial payoffs for each player.
- A is the set of actions available to each player for each pass, which are either tracking ($\mathscr{T}$) or dismissing ($\mathscr{D}$) the pass, so that $A = A_i = \{\mathscr{T}, \mathscr{D}\}\ \forall s_i$.
- J is the payoff function, $J_i(P')$ defined for player s_i and schedule P', which is the sum of the weights of the passes in that schedule that are associated to s_i.
- I_{UW} is the information available to the players at each level of play, which includes the complete set P_{nw}, the set $P_i^s \subset P$ of passes associated to player s_i and the history of play (the Cartesian product of the actions $a \in A$ for each pass from the start to the current level of play). So for each s_i and for each level of play l we have $I_i(l) = \{P_{nw}, P_i^s, H_l\} : p_j \in P_i^s \Leftrightarrow s(p_j) = s_i, \forall a_{p_l} \in A, \forall p_l \in \langle P_{nw}\rangle_{t_s}, H_l = a_{p_1} \times a_{p_2} \times \cdots \times a_{p_l}$.

Note the only information common to all the players is the set of passes with no-priorities P_{nw}, and the history of play H_l. The game differs from the one described in Definition 11 only in that the players do not have information on the priorities of the passes associated to other players.

In this situation the players must plan for the worst case behavior of the rest, that is, considering that for player s_i, other players $s_k : k \neq i$ are aiming at minimizing the payoff of s_i. Before the game starts the players can calculate the best payoff they can achieve for the worst case. Since the history of play is known by all the players, the security strategy must be recomputed as deviations from the security schedule are detected by the players.

Definition 14. The *security schedule of the SRS Game with uncertain priorities for player* s_i and given the history of play H_l is the subset $P_{i,l}^{sm} \subset P$ such that $\forall j \leqslant l$, $p_j \in P_{i,l}^{sm} \Leftrightarrow a_{p_j} = \mathscr{T}$ and $\forall j > l$ no player s_k (with $k \neq i$) unilaterally changing its set of actions can reach an alternative schedule $P' \subset P$ such that $J_i(P') < J_i(P_{i,l}^{sm})$, and player s_i cannot unilaterally reach an alternative schedule $P'' \subset P$ such that $J_i(P'') > J_i(P_{i,l}^{sm})$.

The value $J(P_{i,l}^{sm})$ is the *maximin payoff* for player s_i and history H_l.

The algorithm for finding the security schedule can be obtained easily from the algorithm presented in § 5.3.1.

Corollary 5.1. *The security schedule of a satellite for the SRS game with uncertain priorities can be obtained in* $O(N(\log N)(k_1 + 1)^{k_2})$ *by an extension of the algorithm* *in §5.3.1, where k_1 is the number of resources and (the fixed number) k_2 is the number of players, and N is the number of passes.*

Proof. The algorithm for finding the security schedule is similar to that for the perfect information case, but in this case the way the paths would merge is modified (end time event stages).

Events are generated as in § 5.3.1.1. Start time event nodes creation is performed similarly as explained in § 5.3.1.3. For end time event stages, the last paragraph of (5.3.11) needs to be modified as follows for player s_k to reflect the worst case, that is, all the satellites but s_k aim at minimizing the $J_k(P')$, s_k aims at maximizing $J_k(P')$:

$$\exists^* n_l \in Z_i : n_l \triangleq n_y,$$

$$\text{if} \begin{cases} s_k = s_{\text{lead}} \ \wedge\ w_j(s_k) \geqslant w_y(s_k), & \exists^* v \triangleq (n_l, n_j, w_j), & \exists^* m_l \triangleq m_j, \\ s_k = s_{\text{lead}} \ \wedge\ w_j(s_k) < w_y(s_k), & \exists^* v \triangleq (n_l, n_y, w_y), & \exists^* m_l \triangleq m_y, \quad (5.4.1) \\ s_k \neq s_{\text{lead}} \ \wedge\ w_j(s_k) > w_y(s_k), & \exists^* v \triangleq (n_l, n_y, w_y), & \exists^* m_l \triangleq m_y, \\ s_k \neq s_{\text{lead}} \ \wedge\ w_j(s_k) \leqslant w_y(s_k), & \exists^* v \triangleq (n_l, n_j, w_j), & \exists^* m_l \triangleq m_j. \end{cases}$$

That is, if the evaluated satellite s_k is the current leader, then we take the path with the highest payoff for her; otherwise we take the one with the lowest one. This corresponds to the worst case payoff for s_k, and therefore backtracking the followed path yields the security schedule.

Since the modified algorithm is similar to that from § 5.3.1 except for the criterion for edge creation at end time stages, according to Proposition 5.2 we have that its complexity is in $O(N(\log N)(k_1 + 1)^{k_2})$. □

Note that we followed the same approach as in the modified algorithm regarding cases with the same payoff (5.4.1), and that the algorithm has to be run for every action to be performed where the history of play differs from the previously computed security schedule. Then it is easy to see that if the executed schedule could be calculated in a centralized manner (knowing P), it would take $O(N^2(\log N)(k_1 + 1)^{k_2})$.

Looking at the application of this algorithm in the perfect information case, it is easy to see that the players following this algorithm guarantee a minimum payoff regardless of the opponents' decisions, but these payoffs will by definition be lower than those of the Stackelberg equilibrium.

Figure 5.3 represents the relations between the algorithm providing the optimal solution for the no-slack problem $\mathfrak{R} \mid \tau_j, \overline{\overline{p}}_{ij}, C_\Sigma \mid \sum \mathfrak{w}_j \mathfrak{U}_j$ (§ 3.3) in § 4.2, and the extended algorithms for solving the perfect information $(S, P, A, J, I_{\text{PI}})$ (§ 5.3.1) and uncertain priorities $(S, P, P_{\text{nw}}, A, J, I_{\text{UW}})$ (§ 5.4.1) cases. For each of the dotted line blocks the additions compared to the referenced algorithm are indicated.

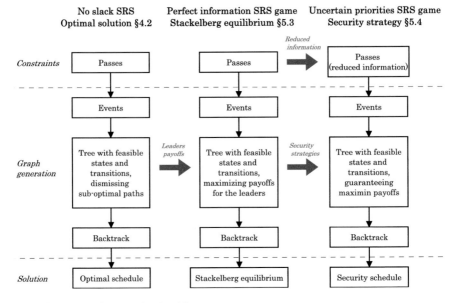

Fig. 5.3 Relations between the algorithms

5.4.2 SRS Game with Uncertain Passes

We now consider the case where the players have no information on the scenario.

Definition 15. The *SRS game with uncertain passes* is the extensive form game $(S, P, A, J, I_{\text{UP}})$ where:

- S is the set of players $\{s_i\}$ (satellites).
- P is the set of passes.
- A is the set of actions available to each player for each pass, which are either tracking ($\mathcal{T}$) or dismissing ($\mathcal{D}$) the pass, so that $A = A_i = \{\mathcal{T}, \mathcal{D}\} \ \forall s_i$.
- J is the payoff function, $J_i(P')$ defined for player s_i and schedule P', which is the sum of the weights of the passes in that schedule that are associated to s_i.
- I_{UP} is the information available to the players at each level of play, which includes the set $P_i^{\text{s}} \subset P$ of passes associated to player s_i, and the history of play of that satellite (the Cartesian product of the actions $a \in A$ for each pass from the start to the current level of play). So for each s_i and for each level of play l we have $I_i(l) = \{P_i^{\text{s}}, H_{i,l}^{\text{s}}\} : p_j \in P_i^{\text{s}} \Leftrightarrow s(p_j) = s_i, \ H_{i,l}^{\text{s}} = a_{p_1} \times a_{p_2} \times \cdots \times a_{p_l} : a_{p_l} \in A, \ \forall p_l \in \langle P_i^{\text{s}} \rangle_{t_s}$.

In this case it is easy to see, given the lack of information about the state of other satellites, that the security strategy is attempting to track every pass.

Proposition 5.3. *The security strategy of a satellite for the SRS game with uncertain passes is attempting to communicate for every single pass.*

Proof. Let us consider a pass $p_j \in P_i^s$ for satellite s_i that is not conflicting with any pass in P_i^s. Since this pass will not avoid any other pass in P_i^s with higher priority to be in the final schedule, the pass p_j will be in the security schedule.

Now let us consider that the pass p_j is conflicting with a pass $p_k \in P_i^s$ with later start time than p_j. If p_j is not in the security schedule, there is the possibility that there is a pass p_l associated to a different satellite and with a start time between those of p_j and p_k, so that p_j will be included in the security schedule.

Therefore the security schedule for player s_i is $P_i^{sm} = P_i^s$.   □

Note however that in this case it is not possible to obtain the maximin payoff, and that its lower bound is zero. Also note that we assume that the set P_i^s is sorted by increasing start time, and for passes with the same start time, by decreasing priority.

5.5 Remarks on the SRS Game

In this section we provide some comments on the approach followed in this chapter.

5.5.1 Equilibrium vs. Security Strategy

Playing *security strategies* guarantees a minimum metric on the schedule of the player. However, in cases with perfect information and rational players, which are the general assumptions for the main case treated in this chapter, the players aim at maximizing their own metrics and they know all the history of play, so they will arrive to an *equilibrium* solution where no one wants to deviate.

5.5.2 Stackelberg Equilibrium vs. Nash Equilibrium

Stackelberg games are characterized by their sequential development, in which the players take turns in a deterministic sequence, assuming the roles of leader and follower. In feedback Stackelberg games, the leader announces her strategy at the beginning of each level of play. The leader knows that the follower will always play his best strategy (depending on the leader's strategy), so that knowing this, and assuming that the players are rational, the leader will play that strategy which maximizes her payoff [5].

Several references on selfish scheduling pursue the calculation of the *Nash equilibrium*, but because of the conflicts among passes the applicability of this equilibrium in our model is found to be limited. If the players selected their strategies at the beginning, they could be selecting passes that will not be tracked because a conflicting pass with earlier start time is being tracked, and information on the conflicts of the passes would be received along the execution of the schedule.

5.5.3 Social Welfare and Price of Anarchy

We present two metrics widely used in the game theory literature for comparing the metric of the solutions of the game and the reference optimal solution. The *social welfare (SW)* is the centralized metric of the obtained schedule $\|P(m')\|_{\Sigma w}$ (5.3.8) and the *price of anarchy (PoA)* is the ratio between the social welfare and the centralized metric of the optimal schedule (§ 4.2), that is $\|P(m')\|_{\Sigma w} / \|P^*\|_{\Sigma w}$. The price of anarchy can be understood as the decrease in performance that mission operators will pay for scheduling selfishly. By definition the upper bound for the social welfare will be the centralized metric of the optimal solution, and the upper bound for the price of anarchy will be one. Their actual values depend on the scenario. We provide some numerical results in § 5.7.

5.5.4 Machine Scheduling vs. SRS

Existing literature on the makespan minimization problem for machine scheduling generally considers the jobs as the players [1–3] (and the selection of the machines as the available actions), so applying this approach to SRS would take to group the passes in coalitions depending on the satellite assigned to them. We dismiss this approach since we are attempting to tackle the problem through noncooperative game theory, although the relations between the Nash equilibrium and the optimal solution of the problem shown in [1] claim for future research in cooperative game theory for this problem following this line.

Current research for game-theoretic approaches in unrelated machine scheduling is focused on the makespan minimization problem [6], and game-theoretic unrelated machine scheduling with deadlines is to the best of our knowledge mostly a barely explored problem. Following the relations presented in Chap. 3, the results in this chapter for SRS could be migrated to the general machine scheduling problem.

5.6 Graph Generation Example

This example aims to explain the algorithms described in this chapter in a simple scenario entailing ground stations g_1 and g_2, satellites s_1 and s_2, and four passes (one for each pair station-satellite). The pass intervals are represented in Fig. 5.4. As previously considered the satellites will be the players of the game.

Fig. 5.4 Passes for the SRS game example

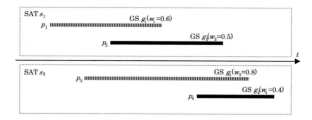

We describe the step-by-step procedure for the creation of the graph in the following lines. For simplicity we will only include the nodes associated to start times of passes in the sub-paths.

The list of the four passes is $P = \{p_1, p_2, p_3, p_4\}$ (so that $\langle P \rangle_{t_s} = \{p_1, p_3, p_2, p_4\}$) which we extend into the set of events $\{e_1, e_2, \ldots, e_8\}$:

$$p_1 = (s_1, g_1, t_1, t_4, w_1) \begin{cases} e_1 = (t_1, +1, s_1, g_1, w_1), \\ e_2 = (t_4, -1, s_1, g_1, 0), \end{cases}$$

$$p_2 = (s_1, g_2, t_3, t_6, w_2) \begin{cases} e_3 = (t_3, +1, s_1, g_2, w_2), \\ e_4 = (t_6, -1, s_1, g_2, 0), \end{cases}$$

$$p_3 = (s_2, g_1, t_2, t_7, w_3) \begin{cases} e_5 = (t_2, +1, s_2, g_1, w_3), \\ e_6 = (t_7, -1, s_2, g_1, 0), \end{cases}$$

$$p_4 = (s_2, g_2, t_5, t_8, w_4) \begin{cases} e_7 = (t_5, +1, s_2, g_2, w_4), \\ e_8 = (t_8, -1, s_2, g_2, 0), \end{cases}$$

where $w_1 = 0.6$, $w_2 = 0.5$, $w_3 = 0.8$ and $w_4 = 0.4$.

We obtain the set E by sorting the events by increasing time, so that:

$$E = \{e_1, e_5, e_3, e_2, e_7, e_4, e_6, e_8\}. \tag{5.6.1}$$

Stage Z_0

For the initialization of the algorithm we have $i = 0$, $B_{-1} = \emptyset$, $n_0 = (0, 0)$, $m_0 = \emptyset$, $v_0 = (n_0, \emptyset, (0, 0))$, and $Z_0 = B_0 = \{n_0\}$.

Extensive form Representation The node n_0 represents the initial vertex in the extensive form representation of the game.

Stage Z_1

The first element in the set E is e_1, which is a start time event. The only node in the frontier B_0 is n_0, so we evaluate it for extension. This node complies with the

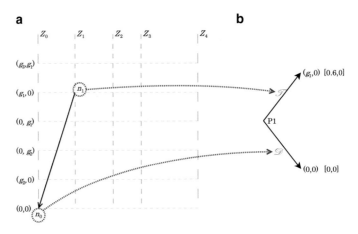

Fig. 5.5 Stage Z_1. (**a**) Graph generation. (**b**) Extensive form representation

conditions stated in first paragraph of (5.3.9): *(a)* the satellite indicated in the event is idle ($g(n_0, s(e_1)) = g(n_0, s_1) = 0$), *(b)* the ground station indicated in the event is idle too ($g(n_0, s') \neq g(e_1) \forall s'$), and *(c)* we have the edge $v_0 = (n_0, \emptyset, (0, 0))$, so that we proceed to create the new node n_1.

From the second paragraph of (5.3.9), the evaluated node is duplicated and the satellite indicated in the event $s(e_1) = s_1$, which was idle, is changed so that it is assigned to the ground station indicated in the event $g(e_1) = g_1$. Therefore the new node will be $n_1 = (g_1, 0)$, and the history vector for this node $m_1 = \{n_1\}$. We also create the new edge from the new node (n_1) to the examined one (n_0), with a payoff for s_1 equal to the sum of the edge from n_0 to $\emptyset$ and the weight of the event $w(e_1) = (w_1, 0)$, so that $v_1 = (n_1, n_0, (w_1, 0))$.

Since there are no more nodes to examine in B_0, we have that the new stage is $Z_1 = \{n_1\}$, and the new frontier $B_1 = B_0 \cup Z_1 = \{n_0, n_1\}$.

EFR The node n_1 corresponds to s_1 playing $\mathcal{T}$, and the node n_0 now represents player 1 playing $\mathcal{D}$. These associations are displayed in Fig. 5.5.

Stage Z_2

We continue with the next event in the set E, the start time event e_5. The nodes to be evaluated in B_1 are n_0 and n_1. The node n_1 cannot be extended since the ground station indicated in the event is not idle ($g(e_5) = g(n_1, s_1) = g_1$).

Then only the node n_0 can be extended, which following the procedure in the previous stage yields a new node $n_2 = (0, g_1)$, with a history vector $m_2 = \{n_2\}$ and an edge $v_2 = (n_2, n_0, (0, w_3))$. Then, $Z_2 = \{n_2\}$, and $B_2 = B_1 \cup Z_2 = \{n_0, n_1, n_2\}$.

EFR The sub-tree from the play $\{\mathscr{T},\mathscr{T}\}$ is unfeasible, which we represent starting with a red cross in the extensive form. Plays $\{\mathscr{D},\mathscr{D}\}$, $\{\mathscr{T},\mathscr{D}\}$ and $\{\mathscr{D},\mathscr{T}\}$ correspond to the nodes n_0, n_1, and n_2, respectively, as shown in Fig. 5.6.

Stage Z_3

The next event in E is e_3, which is also an start time event, so we follow the previous procedure to evaluate nodes n_0, n_1, and n_2. Node n_1 is dismissed since the satellite $s(e_3) = s_1$ is not idle ($g(n_1, s(e_3)) = g_1 \neq 0$). Extending nodes n_0 and n_2 we create $n_3 = (g_2, 0)$, $m_3 = \{n_3\}$, $v_3 = (n_3, n_0, (w_2, 0))$ and $n_4 = (g_2, g_1)$, $m_4 = \{n_2, n_4\}$, $v_4 = (n_4, n_2, (w_2, w_3))$, so that $Z_3 = \{n_3, n_4\}$ and $B_3 = \{n_0, n_1, n_2, n_3, n_4\}$.

EFR Play $\{\mathscr{T},\mathscr{D},\mathscr{T}\}$ is unfeasible. Plays $\{\mathscr{D},\mathscr{D},\mathscr{D}\}$, $\{\mathscr{T},\mathscr{D},\mathscr{D}\}$, $\{\mathscr{D},\mathscr{T},\mathscr{D}\}$, $\{\mathscr{D},\mathscr{D},\mathscr{T}\}$ and $\{\mathscr{D},\mathscr{T},\mathscr{T}\}$ correspond to nodes n_0, n_1, n_2, n_3, and n_4, respectively. These relations are shown in Fig. 5.7.

Stage Z_4

We take now the next event in E, which is the end time event e_2 ($\phi(e_2) < 0$). Checking the conditions in (5.3.11), we only evaluate the nodes in B_3 that have the ground station $g(e_2) = g_1$ assigned to the satellite $s(e_2) = s_1$, that is, the node $n_1 = (g_1, 0)$. We have that $m_1 = \{n_1\}$ and $v_1 = (n_1, n_0, (w_1, 0))$, and also there is only a node in B_3 that has the same state that n_1 except for the satellite $s(e_2) = s_1$, which is idle. This node is $n_0 = (0, 0)$, with $m_0 = \emptyset$ and $v_0 = (n_0, \emptyset, (0, 0))$.

The comparison of m_0 and m_1 yields that the leader satellite is that one associated to n_1, which is s_1 (pass p_1). This satellite has a higher payoff in n_1, so that we create then the new node $n_5 = (0, 0)$, $m_5 = m_1 = \{n_1\}$ and the new edge $v_5 = (n_5, n_1, (w_1, 0))$. As no more nodes from B_3 can be extended, the set of created nodes is $\{n_5\}$, and the set of deleted nodes is $\{n_0, n_1\}$. Therefore the new frontier is $\{B_3 \cup \{n_5\}\} - \{n_0, n_1\}$, which is $B_4 = \{n_2, n_3, n_4, n_5\}$.

EFR The plays $\{\mathscr{T},\mathscr{D},\mathscr{D}\}$ and $\{\mathscr{D},\mathscr{D},\mathscr{D}\}$ would converge to the same state $(0, 0)$, and generate two sub-trees respectively corresponding to strategically equivalent games. The leading player in this case is s_1, which prefers the sub-tree associated to the play $\{\mathscr{T},\mathscr{D},\mathscr{D}\}$ where it gets a better payoff, and thus we can delete the other sub-tree, as shown in Fig. 5.8.

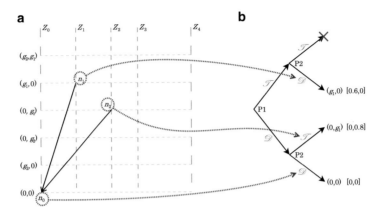

Fig. 5.6 Stage Z_2. (**a**) Graph generation. (**b**) Extensive form representation

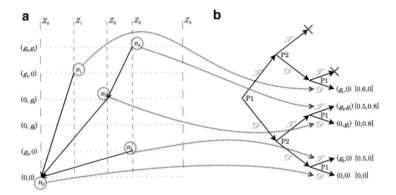

Fig. 5.7 Stage Z_3. (**a**) Graph generation. (**b**) Extensive form representation

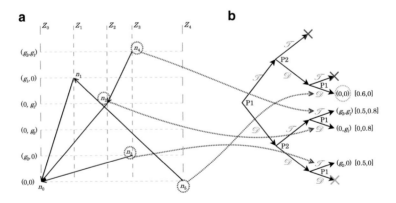

Fig. 5.8 Stage Z_4. (**a**) Graph generation. (**b**) Extensive form representation

Rest of Stages

We continue with the execution of the algorithm through the rest of the events in E, generating the rest of nodes, edges, and history vectors (with only start time nodes):

$$
\begin{aligned}
n_0 &= (0,0), & v_0 &= (n_0, \emptyset, (0,0)), & m_0 &= \emptyset, \\
n_1 &= (g_1, 0), & v_1 &= (n_1, n_0, (.6, 0)), & m_1 &= \{n_1\}, \\
n_2 &= (0, g_1), & v_2 &= (n_2, n_0, (0, .8)), & m_2 &= \{n_2\}, \\
n_3 &= (g_2, 0), & v_3 &= (n_3, n_0, (.5, 0)), & m_3 &= \{n_3\}, \\
n_4 &= (g_2, g_1), & v_4 &= (n_4, n_2, (.5, .8)), & m_4 &= \{n_2, n_4\}, \\
n_5 &= (0, 0), & v_5 &= (n_5, n_1, (.6, 0)), & m_5 &= \{n_1\}, \\
n_6 &= (0, g_2), & v_6 &= (n_6, n_5, (.6, .4)), & m_6 &= \{n_1, n_6\}, \\
n_7 &= (0, 0), & v_7 &= (n_7, n_5, (.6, 0)), & m_7 &= \{n_1\}, \\
n_8 &= (0, g_1), & v_8 &= (n_8, n_4, (.5, .8)), & m_8 &= \{n_2, n_4\}, \\
n_9 &= (0, 0), & v_9 &= (n_9, n_7, (.6, 0)), & m_9 &= \{n_1\}, \\
n_{10} &= (0, 0), & v_{10} &= (n_{10}, n_6, (.6, .4)), & m_{10} &= \{n_1, n_6\},
\end{aligned}
$$

and the stages and frontiers:

$$
\begin{aligned}
Z_0 &= \{n_0\}, & B_0 &= \{n_0\}, \\
Z_1 &= \{n_1\}, & B_1 &= \{n_0, n_1\}, \\
Z_2 &= \{n_2\}, & B_2 &= \{n_0, n_1, n_2\}, \\
Z_3 &= \{n_3, n_4\}, & B_3 &= \{n_0, n_1, n_2, n_3, n_4\}, \\
Z_4 &= \{n_5\}, & B_4 &= \{n_2, n_3, n_4, n_5\}, \\
Z_5 &= \{n_6\}, & B_5 &= \{n_2, n_3, n_4, n_5, n_6\}, \\
Z_6 &= \{n_7, n_8\}, & B_6 &= \{n_6, n_7, n_8\}, \\
Z_7 &= \{n_9\}, & B_7 &= \{n_6, n_9\}, \\
Z_8 &= \{n_{10}\}, & B_8 &= \{n_{10}\}.
\end{aligned}
$$

Figure 5.9 shows the passes at the top, the algorithm graph at the middle, and the game in extensive form at the bottom. For the extensive form representation, infeasible sub-trees start with a red cross and are gray-colored; and dismissed sub-trees from the leader players start with a green cross [and are connected with a curved arrow to the other examined node, nodes n_j and n_y in (5.3.11)].

Also, payoff vectors are represented with brackets to differentiate them from the state vectors (with equilibrium solution payoffs in bold letters). We consider that tracking an unfeasible pass subtracts from the deciding player an arbitrary payoff of $\omega = 10$ (note that from Proposition 5.1 a lower value would be enough for guaranteeing a negative payoff).

Those edges that have not been created (but checked for creation) have been represented as light gray dotted lines, and the backtracked path is in bold lines. The equilibrium schedule can be obtained by taking the elements in the last created history vector m_{10}, so that $P(m_{10}) = \{p_1, p_4\}$.

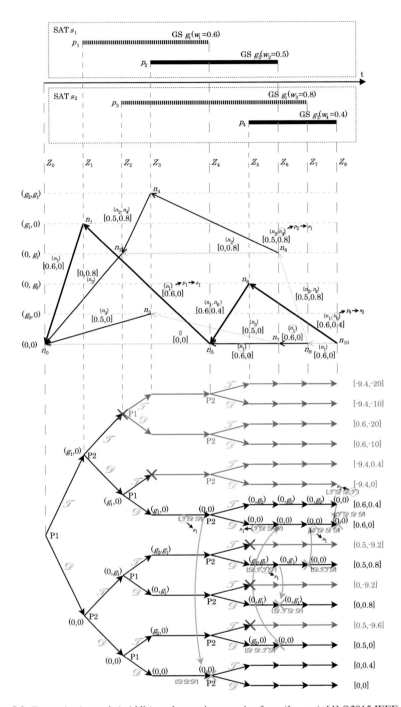

Fig. 5.9 Passes (*top*), graph (*middle*), and game in extensive form (*bottom*). [4] ©2015 IEEE

The main difference compared to the referenced algorithm can be seen in the transition from stage Z_6 to Z_7, where the agent that decides is the satellite 1 because the difference between the merging paths is the first pass, which corresponds to s_1, for which it has a higher payoff.

Although the optimal schedule would be $P^* = \{p_2, p_3\}$, with a centralized metric of 1.3, the Stackelberg equilibrium yields a centralized metric of 1.0, so that the price of anarchy for this example is 0.77.

5.7 Simulations

In this section we present some results for comparing the equilibrium to reference solutions in practical scenarios. As for the presented example, we focus on the problem with perfect information.

The simulations were run in MATLAB on a virtual machine with a 3 GHz processor and 2 GB RAM, and we followed the relaxation given after the proof of Proposition 5.2 for minimizing the space complexity of the implementation.

We use the two same scenarios as in Chap. 4 (§ 4.6). Both scenarios have $|G| = 5$, $|S| = 5$, a scheduling horizon ranging from 1 to 14 days, and fixed interval passes with random priorities $w(p_l) = v_l/10 : v_l \sim U[1,10]$. Whereas for scenario 1 we assign arbitrary positions to the ground stations, and consider satellites in different low Earth orbits (LEO), scenario 2 corresponds to the worst case scenario for the presented algorithm, wherein all the locations are identical, and so are all the orbits; so that this case yields the maximum number of conflicts.

The performance metrics we are using are the *social welfare (SW)* and the *price of anarchy (PoA)*, which we introduced in § 5.5.

Strategies for the players can be either *optimal* (§ 4.2), *greedy earliest start time* (see, for example, [7]) or *selfish* (§ 5.3). We consider five cases with different combinations of these strategies:

Case 1: Optimal   All the satellites follow the optimal schedule, therefore arriving to the optimal solution, which corresponds to the maximum social welfare. This solution can be calculated through the algorithm presented in § 4.2. Note however that these players are not rational, because they are not taking actions to maximize their associated payoffs. In fact, from Definition 12, *any strategy different from the selfish one is not rational.*

Case 2: Selfish vs. optimal   We modify the previous case by introducing a selfish player in the set of optimal players. Assuming perfect information, we consider that this selfish player knows that the other players are following the optimal schedule. This requires to modify the strategy of this player for taking into account the strategies of the other players (note that the algorithm provided in § 5.3.1 considers that all the players are rational). Whereas this player will unilaterally improve its payoff by deviating from the optimal strategy, the others will not change their schedules regardless of whether their passes are actually tracked or not. For the implementation, we first obtain the optimal schedule, then we add to that list of passes all those in P associated to the selfish satellite, and then we apply the algorithm in § 5.3.1.

Case 3: Greedy (earliest start time)   In this case all the satellites greedily select the next pass that is available to be tracked (sorted by increasing start time), arriving to a suboptimal solution. As we introduced previously, the most used heuristics for greedy algorithms are earliest start time [7], earliest deadline [8], or priority [9]. In this case we consider the earliest start time heuristic, which we consider the most simple one for the extensive formulation of the game.

Case 4: Selfish vs. greedy   We introduce a selfish player which knows that the other players are playing greedy strategies. For the implementation, we modify the algorithm so that in (5.3.11), if the leader satellite is greedy, the pass is tracked regardless of the associated payoff; whereas if the leader is the selfish satellite, it still selects the path with highest payoff.

Case 5: Selfish   Finally we consider the case where all the players act selfishly, arriving to the Stackelberg equilibrium. We have already shown that this equilibrium can be computed following the algorithm in § 5.3.1. Even though this solution is suboptimal, no rational player has an incentive to deviate from it as we proved in Proposition 5.1.

We implemented these five cases for the two presented scenarios, which execution times we represent in Figs. 5.10a and 5.11a. We also show the price of anarchy versus the number of passes in P for T ranging from 1 to 14 days for the two scenarios in Figs. 5.10b and 5.11b, respectively.

These results show that the performance of the presented algorithm (*case 5*) is lower than that of the optimal (*case 1*) both in terms of execution time (complexity of the algorithm) and in terms of metric (social welfare), and if so, why would the

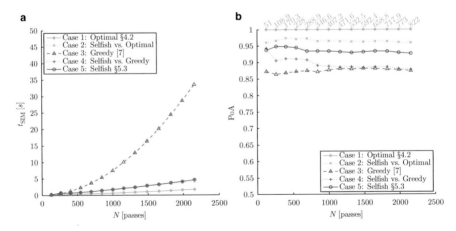

Fig. 5.10 Simulation results: scenario 1. (**a**) Simulation times. (**b**) Price of anarchy. [4] ©2015 IEEE

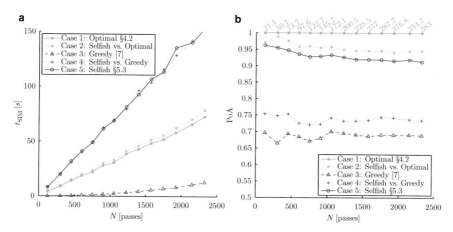

Fig. 5.11 Simulation results: scenario 2. (**a**) Simulation times. (**b**) Price of anarchy. [4] ©2015 IEEE

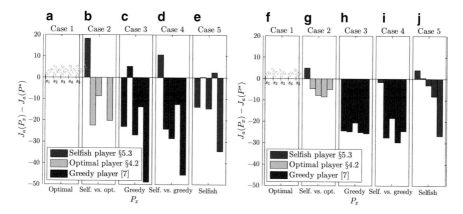

Fig. 5.12 Payoff differences. (**a**)–(**e**) Scenario 1 results. (**f**)–(**j**) Scenario 2 results. [4] ©2015 IEEE

agents want to deviate? The answer to this question is on the selfish behavior of the players, which do not care on the centralized metric but on their own payoffs.

We compare the payoffs for the five players for each of the cases for $T = 14$ days. We represent these results in Fig. 5.12 for scenario 1 (left graph) and scenario 2 (right graph). In both graphs we show groups of differences in payoffs for the studied case compared to the optimal case. That is, for each case and for each satellite we show the value of $J_a(P_x) - J_a(P^*)$ with $s_a = \{s_1, s_2, \ldots, s_5\}$ versus P_x, which is the schedule resulting from running the cases described in previous lines, so that $x = 1, 2, \ldots, 5$ corresponds to the studied case. Each of these graphs is divided into subgraphs corresponding to the studied cases.

We provide more detail about these results in the following lines:

Case 1: Optimal   It is easy to see that for this case the payoff differences are null for all the satellites, that is $J_a(P_1) - J_a(P^*) = 0 \ \forall a$, since $P_1 = P^*$. We display the values for the payoffs of the players ($\{s_1, s_2, \ldots, s_5\}$) playing *optimal* strategies in Fig. 5.12a, f for scenarios 1 and 2 respectively.

Case 2: Selfish vs. optimal   If we introduce a selfish player s_1 in the pool of optimal players, this player will improve its associated payoff causing losses to the other players. Results are displayed in Fig. 5.12b, g.

Case 3: Greedy (earliest start time)   Figure 5.12c, h show that on average the greedy players get lower payoffs than those corresponding to the optimal solution.

Case 4: Selfish vs. greedy   If we introduce a selfish player s_1 in a group of greedy players, again the selfish player increases unilaterally its payoff. These results are displayed in Fig. 5.12d, i. Note that for the scenario 2 (Fig. 5.12i), the payoff difference for the selfish player is negative (that is $J_1(P_4) < J_1(P_1)$), but it is in fact positive compared to its playing the greedy strategy ($J_1(P_4) > J_1(P_3)$).

Case 5: Selfish   The payoffs for all the players playing selfishly are represented in Fig. 5.12e, j. The solution is suboptimal, but the schedule is stable in the sense that no player can get a better payoff unilaterally.

5.8 Summary

We extend the notation in Chap. 3 (§ 3.3) to cover the problems presented in this chapter. We add the symbol $\mathfrak{D}$ for representing the case with several unrelated entities with independent schedulers, that is, the distributed version of $\mathfrak{R}$.

- *Perfect information*: if the first term is $\mathfrak{D}$, and the notation includes no additional symbols, we assume that the problem has perfect information.
- *Uncertain priorities*: in order to indicate that the priorities of the passes of other players are not known, we add the wide hat symbol to the terms in the optimization function indicating the priorities, that is, $\sum \hat{w}_j \mathfrak{U}_j$.
- *Uncertain history*: in order to indicate that the history of the plays of other players is not known, we add the wide hat symbol to the term in the optimization function indicating if the pass is in the schedule, that is, $\sum w_j \hat{\mathfrak{U}}_j$.
- *Uncertain passes*: in order to indicate that the players have no information about other players we add the wide hat symbol to the first term, that is, $\hat{\mathfrak{D}}$.

We summarize the problems and proposed game-theoretic models in Table 5.1, classifying them according to the information available to the players (info. stands for information). For all the problems we assume multiple resources, priorities, no-preemption, no-precedence, and no-slack.

We show the relations among these problems in Fig. 5.13. Intermediate information versions of the problem are included for completeness. The dashed arrows

Table 5.1 Noncooperative SRS problems

Problem	Info. passes		Info. priorities		History		Game model	Game solution	
	Yes	No	Yes	No	Yes	No		Stackelberg	Security
$\mathfrak{D}\,\vert\,\tau_j,\overline{\overline{p_{ij}}},C_\Sigma\,\vert\,\sum w_j\mathfrak{U}_j$	X		X		X		$(S,P,A,J,I_{\mathrm{PI}})$	§ 5.3.1	N/A
$\mathfrak{D}\,\vert\,\tau_j,\overline{p_{ij}},C_\Sigma\,\vert\,\sum \hat{w}_j\mathfrak{U}_j$	X			X	X		$(S,P,A,J,I_{\mathrm{UW}})$	–	§ 5.4.1
$\mathfrak{D}\,\vert\,\tau_j,\overline{p_{ij}},C_\Sigma\,\vert\,\sum w_j\hat{\mathfrak{U}}_j$	X		X			X	Open problem	–	–
$\mathfrak{D}\,\vert\,\tau_j,\overline{p_{ij}},C_\Sigma\,\vert\,\sum \hat{w}_j\hat{\mathfrak{U}}_j$	X			X		X	Open problem	–	–
$\hat{\mathfrak{D}}\,\vert\,\tau_j,\overline{p_{ij}},C_\Sigma\,\vert\,\sum w_j\mathfrak{U}_j$		X		X		X	$(S,P,A,J,I_{\mathrm{UP}})$	N/A	§ 5.4.2

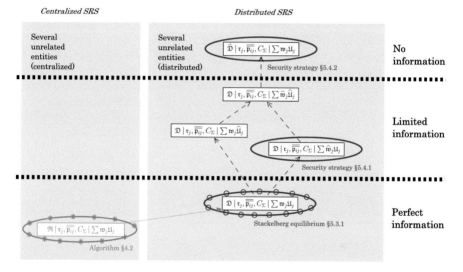

Fig. 5.13 Noncooperative SRS and relations with centralized SRS

among the distributed SRS problems represent reductions on the information available to the players.

Acknowledgements This research was performed while the author held a National Research Council Research Associateship Award at the Air Force Research Laboratory (AFRL).

References

1. C. Wang, Y. Xi, Noncooperative game theory as a unified framework for single machine scheduling, in 5th *Asian Control Conference* (IEEE, Melbourne, 2004)
2. N. Immorlica, L. Li, V.S. Mirrokni, A. Schulz, Coordination mechanisms for selfish scheduling, in *Workshop on Internet and Network Economics* (Springer, Heidelberg, 2005)
3. L. Agussurja, H.C. Lau, The price of stability in selfish scheduling games, in *International Conference on Intelligent Agent Technology* (IEEE, Fremont, 2007)

4. A.J. Vazquez, R.S. Erwin, Noncooperative satellite range scheduling with perfect information, in *2015 IEEE Aerospace Conference* (IEEE, Big Sky, 2015)
5. T. Basar, G.J. Olsder, *Dynamic Noncooperative Game Theory*. Classics in Applied Mathematics (SIAM, Philadelphia, 1982)
6. B. Heydenreich, R. Müller, M. Uetz, Games and mechanism design in machine scheduling - an introduction. Prod. Oper. Manag. **16**(4), 437–454 (2007)
7. A. Globus, J. Crawford, J. Lohn, A. Pryor, A comparison of techniques for scheduling Earth observing satellites, in *Proceedings of the Sixteenth Innovative Applications of Artificial Intelligence Conference* (IAAI, San Jose, 2004)
8. L. Barbulescu, J.P. Watson, L.D. Whitley, A.E. Howe, Scheduling space-ground communications for the Air Force Satellite Control Network. J. Sched. **7**(1), 7–34 (2004)
9. W.J. Wolfe, S.E. Sorensen, Three scheduling algorithms applied to the Earth observing systems domain. Manag. Sci. **46**(1), 148–168 (2000)

Chapter 6
Robust Satellite Range Scheduling

In this chapter we introduce nondeterministic uncertainty modeling into the scenario. We will focus on the case where passes on the schedule could be dropped from the final schedule with a certain probability, modeling the associated satellite or ground station failing to establish communication due to factors not included or that changed from what was included in the scheduling problem (weather, maintenance, etc.). Through the presented framework we will also tackle the versions of the problem with uncertainty on the priorities of the passes, and on their duration.

Existing literature has previously tackled the single scheduling entity version of this problem through different suboptimal approaches, for example, based on fuzzy logic [1] or adding backup passes to an initial schedule [2]. We model uncertainty through the introduction of random variables associated to the passes, so that rather than a deterministic optimal solution we will be looking for the schedule providing the maximal expected metric. Compared to deterministic versions where the space of possible solutions was limited to feasible schedules, the introduction of uncertainty complicates the problem, as in this case considering unfeasible schedules will allow to have additional passes in case others fail. Furthermore, the introduction of priorities will yield a combinatorial explosion of the possible results of a solution.[1]

---

[1] This chapter is strongly based on our work [3], which has been extended, reorganized, and integrated with the rest of the book. For a detailed list of the new contributions please see § 1.5.

© Springer International Publishing Switzerland 2015
A.J. Vázquez Álvarez, R.S. Erwin, *An Introduction to Optimal Satellite Range Scheduling*, Springer Optimization and Its Applications 106,
DOI 10.1007/978-3-319-25409-8_6

6.1 Scenario Model for Robust SRS

We briefly repeat the formulation of the SRS problem for completeness of the chapter. For more details Chaps. 3 and 4 can be consulted.

Let $S = \{s_i\}$ be a set of satellites, and $G = \{g_h\}$ a set of ground stations. As for previous chapters, we consider a scheduling horizon T starting at t_0: $t \in [t_0, t_0 + T]$. We consider that a pass $p_l = (s_i, g_h, t_{s_l}, t_{e_l}, w_l)$ is conflicting with an earlier start time pass p_m if they are time overlapping and both are associated to either the same satellite or ground station (6.1.1).

Remark 6.1. As for previous chapters (§ 3.1.2) we assume no redundancy (a ground station can only communicate to one satellite at the same time and vice-versa).

We say that $P_{\text{sub}} \subset P$ is a *feasible schedule* if all its passes are non-conflicting (6.1.2). Given a feasible schedule P^{f}, let the *metric* $\| \cdot \|_{\Sigma w}$ be the sum of the priorities of the passes in this schedule (6.1.3). The objective of the *SRS problem* is finding a feasible schedule with maximal metric (6.1.4). That is:

$$\phi(p_l, p_m) = 1 \Leftrightarrow \{(g(p_l) = g(p_m)) \vee (s(p_l) = s(p_m))\} \wedge \\ \{t_{s_l} \in [t_{s_m}, t_{e_m}]\} \wedge \{l \neq m\}, \tag{6.1.1}$$

$$P_{\text{sub}} \in \{P^{\text{f}}\} \Leftrightarrow \sum \phi(p_l, p_m) = 0 \ \ \forall p_l, p_m \in P_{\text{sub}}, \tag{6.1.2}$$

$$\|P^{\text{f}}\|_{\Sigma w} = \sum_l w_l \ : \ w_l = w(p_l) \ \forall p_l \in P^{\text{f}}, \tag{6.1.3}$$

$$P^* \triangleq \arg \max(\|P_{\text{sub}}\|_{\Sigma w}) \ \forall P_{\text{sub}} \in \{P^{\text{f}}\}, \tag{6.1.4}$$

where $\{P^{\text{f}}\}$ is the set of feasible schedules. In the remainder of the chapter we will refer to this problem as the *basic SRS problem*.

Note that from Chaps. 3 and 4 this basic problem is deterministic, or equivalently that all the entities are 100 % reliable, so that no pass in the optimal schedule will fail in the future. In this chapter, however, we consider that a pass $p_l \in P$ can be dropped from the schedule with probability α_l.

Let $P' \subset P$ be a schedule, and let $\widetilde{P'} \subset P'$ be the *executed schedule* of P' (or also, a realization of P'). That is, $\widetilde{P'}$ is the set of passes that were not dropped from P' at the end of the scheduling horizon $t_0 + T$, either because of the failure probabilities or earlier start time conflicting passes being in this set. In summary, P would be the initial set of passes (problem definition), $P' \subset P$ the selected schedule (proposed solution), and $\widetilde{P'} \subset P'$ the executed schedule (actual result).

The scenario for the robust SRS problem is represented in Fig. 6.1, where the links among the ground stations and the satellites are not reliable.

Remark 6.2. We assume that the failure probabilities α_l are independent, and model all the factors in the system which introduce uncertainty: satellite or ground station related (α_S, α_G), weather related (α_W), etc. Even if we introduced time-dependency

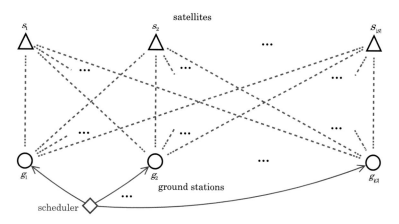

Fig. 6.1 Robust scheduler

(α_t) in the model, the start times of the passes are fixed, and thus the values of α_l would be fixed. Knowledge on the model generating these priorities, e.g., $\alpha_l = f(\alpha_S, \alpha_G, \alpha_W, \alpha_t)$ would simplify the problem. Therefore, assuming independence among these probabilities, given a known P', we have that the probability of a pass p_l to be in the executed schedule $\widetilde{P'}$ given that it is in the schedule P' is:

$$\mathbb{P}(p_l \in \widetilde{P'} \mid p_l \in P') = (1 - \alpha_l) \cdot \mathbb{P}(\nexists p_m \in \widetilde{P'} : \phi(p_l, p_m) = 1). \qquad (6.1.5)$$

From Remark 6.1 we have that the set $\widetilde{P'}$ has to be feasible $\widetilde{P'} \in \{P^f\}$.

Definition 16. We define the *expected metric of a schedule* P' as the expected value of the metric of the executed schedule:

$$\|P'\|_{\mathbb{E}} = \mathbb{E}\{\|\widetilde{P'}\|_{\Sigma w}\} = \sum_{p_l \in P'} w_l \cdot \mathbb{P}(p_l \in \widetilde{P'} \mid p_l \in P'). \qquad (6.1.6)$$

We can now state the objective of the robust satellite range scheduling problem:

Definition 17. The *robust SRS* problem can be stated as finding the *robust schedule* P^R, which is a schedule with maximal expected metric.

$$P^R \subseteq P, \quad \nexists P_{\text{sub}} \subseteq P : \|P_{\text{sub}}\|_{\mathbb{E}} > \|P^R\|_{\mathbb{E}}. \qquad (6.1.7)$$

Note that in contrast to the basic SRS problem we now should not restrict the space of possible solutions $P' \subset P$ to feasible schedules, even though executed schedules are necessarily restricted so. Extending the set of possible solutions to all the possible subsets of P allows to have conflicting passes that, even though they may fail, they could allow later conflicting passes to contribute to the metric of the executed schedule.

6.1.1   Complexity of the Robust SRS Problem

In this section we study the tractability of finding the robust schedule.

Lemma 6.1. *The Robust SRS problem generalizes the Basic SRS problem.*

Proof. If $\alpha_l = 0 \ \forall p_l \in P$, then from (6.1.5) we have that $\mathbb{P}(p_l \in \widetilde{P}' \mid p_l \in P') = \mathbb{P}(\nexists p_m \in \widetilde{P}' : \phi(p_l, p_m) = 1)$, and a pass from P' will be in $\widetilde{P}'$ if there are no prior conflicting passes in $\widetilde{P}'$. Therefore, given any feasible schedule $P' \subset P : P' \in \{P^f\}$, we have $p_l \in P' \Rightarrow p_l \in \widetilde{P}'$, so that we can reduce the space of possible solutions to the space of feasible schedules (since no pass will be dropped), and thus $\|P^R\|_{\mathbb{E}} = \|P^*\|_{\Sigma w}$ if $\alpha_l = 0 \ \forall p_l \in P$. □

Corollary 6.1. *The robust no-slack SRS problem is NP-hard.*

Proof. In Theorem 3.1 (§ 3.2.2) we show that the decision version of the basic SRS problem is NP-complete, and thus the optimization version is NP-hard. From Lemma 6.1, robust SRS has to be at least as complicated as basic SRS, and therefore Robust SRS is NP-hard. □

The problem is simplified however if we consider a single scheduling entity. Let us focus for now on this case, before studying the multiple entity case in § 6.3.

6.2   Restricted Robust SRS Problem

In this section we provide an algorithm for finding the robust schedule for instances of the problem with a single scheduling entity.

We assume that the set P is ordered by increasing start time, i.e., $P = \{p_1, p_2, \ldots, p_N\}$ such that $t_s(p_i) < t_s(p_j) \Leftrightarrow i < j$.

Definition 18. Let the subset $P_l \subset P$ associated to p_l be the set of passes $\{p_l, p_{l+1}, \ldots, p_N\} \subset P$, such that $p_l \in P_l$ and $p_k \in P_l \Leftrightarrow t_s(p_l) < t_s(p_k)$. We define the *set of later non-conflicting passes* for the pass p_l as the subset D_l whose passes have a later start time than p_l and which are not conflicting with it. Then:

$$\forall p_l \in P, \ \exists D_l \subset P_l : \ \forall p_d \in P_l, \ p_d \in D_l \Leftrightarrow \\ t_s(p_l) < t_s(p_d) \wedge \phi(p_d, p_l) = 0, \tag{6.2.1}$$

so that if $\nexists p_d \in P_l : t_s(p_l) < t_s(p_d) \wedge \phi(p_d, p_l) = 0$ then $D_l = \emptyset$.

We provide a result based on this concept of later non-conflicting passes which is applicable to scenarios with a single scheduling entity.

Proposition 6.1. *Let P be a set of passes sorted by increasing start time. If there is a single scheduling entity ($|G| = 1$ or $|S| = 1$), we have that $D_l \cap P_l = P_x \ \forall p_l \in P$, where $p_x \in P_l$ is the pass with earliest start time not conflicting with p_l.*

Proof. If whether $|G| = 1$ or $|S| = 1$, then from (6.1.1):

$$\phi(p_m, p_l) = 1 \Leftrightarrow t_{s_m} \in [t_{s_l}, t_{e_l}] \ \forall p_l, p_m \in P, \ l \neq m, \tag{6.2.2}$$

so that time overlap is a sufficient condition for two passes to be conflicting. Therefore, given a pass $p_l \in P$, if there is a pass p_m with later start time and not conflicting with it, then any pass with later start time than p_m is not conflicting with p_l:

$$\forall p_l, p_m, p_k \in P : \ t_s(p_l) < t_s(p_m) < t_s(p_k), \\ \phi(p_m, p_l) = 0 \Rightarrow \phi(p_k, p_l) = 0. \tag{6.2.3}$$

Let $p_x \in P_l$ be the pass with the earliest start time in D_l, then $\forall p_m \in P_l$ we have that $t_s(p_x) < t_s(p_m) \Rightarrow \phi(p_m, p_l) = 0$, thus $p_m \in D_l$, and therefore $D_l \cap P_l = P_x$. $\quad\square$

Note that this result is not applicable to the case where both $|G|$ and $|S|$ are greater than one. This is illustrated in the following two examples. The first example in Fig. 6.2 shows a set of passes for a single ground station. For pass p_1 it is easy to see that the first non-conflicting pass is p_4, and any pass with later start time is not conflicting with p_1. The second example in Fig. 6.3 shows a set of passes for two ground stations. In this case the first non-conflicting pass for p_1 is p_2 (which although time overlapping, is associated to a different ground station). The pass p_3 has a later start time than p_2, but it is conflicting with p_1.

This property described in Proposition 6.1 for scenarios with a single scheduling entity allows for an iterative calculation of the metric of the schedule:

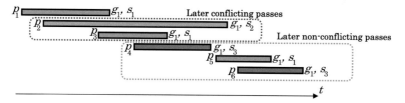

Fig. 6.2 First example: set of later non-conflicting passes for a single scheduling entity

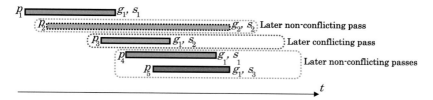

Fig. 6.3 Second example: set of later non-conflicting passes for several scheduling entities

Lemma 6.2. *The expected metric of a schedule* $P' = \{p_y, \ldots, p_z\} \subset P$ *with a single scheduling entity can be computed recursively in* $O(N)$ *as* $\|P'\|_{\mathbb{E}} = \|P'_y\|_{\mathbb{E}}$ *where:*

$$\|P'_l\|_{\mathbb{E}} = \begin{cases} (1 - \alpha_l)(w_l + \|D_l \cap P'_{l+1}\|_{\mathbb{E}}) + \alpha_l \|P'_{l+1}\|_{\mathbb{E}}, & p_l \in P', \\ \|P'_{l+1}\|_{\mathbb{E}}, & p_l \notin P', \end{cases} \quad (6.2.4)$$

for all $p_l \in P'$, *and with* $\|P'_l\|_{\mathbb{E}} = 0 \; \forall l > z$ *for consistency of the algorithm.*

Proof. Let $p_x \in P'$ be the pass with earliest start time that is not conflicting with p_l, such that $t_s(p_l) < t_s(p_x)$, and let $P'_x = P_x \cap P'$ and $P'_l = P_l \cap P'$. Since there is a single scheduling entity, we have from Proposition 6.1 that $D_l \cap P'_l = P'_x$.

We calculate the expected metric of the sub-schedule P'_l. For simplicity of notation we will work with P, P_l, and P_x (instead of P', P'_l, and P'_x), since this will allow us to keep all the indexes for the passes $k = 1, 2, \ldots, N$. There are two possible cases, *(i)* $p_l \in P_l$ and *(ii)* $p_l \notin P_l$, which we explain in the following lines:

(i) If $p_l \in P_l$, from (6.1.6) we have:

$$\|P_l\|_{\mathbb{E}} = \sum_{k=l}^{N} w_k \cdot \mathbb{P}\big(p_k \in \widetilde{P_l} \mid p_k \in P_l, \, p_l \in P_l\big). \quad (6.2.5)$$

Since $p_l \in P_l$, we have that either $p_l \in \widetilde{P_l}$ or $p_l \notin \widetilde{P_l}$, with probabilities $\mathbb{P}(p_l \in \widetilde{P_l} \mid p_l \in P_l) = (1 - \alpha_l)$ and $\mathbb{P}(p_l \notin \widetilde{P_l} \mid p_l \in P_l) = \alpha_l$. Therefore:

$$\mathbb{P}\big(p_k \in \widetilde{P_l} \mid p_k \in P_l, \, p_l \in P_l\big) = (1 - \alpha_l) \cdot \mathbb{P}\big(p_k \in \widetilde{P_l} \mid p_k \in P_l, \, p_l \in \widetilde{P_l}\big) + \\ \alpha_l \cdot \mathbb{P}\big(p_k \in \widetilde{P_l} \mid p_k \in P_l, \, p_l \notin \widetilde{P_l}\big). \quad (6.2.6)$$

We consider three groups of passes p_k: the pass p_l $(k = l)$, the set of later passes conflicting with it $(l + 1 \leqslant k \leqslant x - 1)$, and the rest $(x \leqslant k \leqslant N)$, so that the probabilities in (6.2.6) are:

$$\mathbb{P}\big(p_k \in \widetilde{P_l} \mid p_k \in P_l, \, p_l \in \widetilde{P_l}\big) = \begin{cases} 1, & k = l, \\ 0, & l < k < x, \\ \mathbb{P}\big(p_k \in \widetilde{P_x} \mid p_k \in P_x\big), & x \leqslant k \leqslant N, \end{cases} \quad (6.2.7)$$

$$\mathbb{P}\big(p_k \in \widetilde{P_l} \mid p_k \in P_l, \, p_l \notin \widetilde{P_l}\big) = \begin{cases} 0, & k = l, \\ \mathbb{P}\big(p_k \in \widetilde{P_{l+1}} \mid p_k \in P_{l+1}\big), & l < k \leqslant N. \end{cases} \quad (6.2.8)$$

For consistency, if $D_l = \emptyset$ then we would only consider the cases $k = l$ and $l < k \leqslant N$ (or simply that $x = N + 1$) in (6.2.7). Developing (6.2.5) throughout (6.2.6)–(6.2.8):

$$\|P_l\|_{\mathbb{E}} = (1 - \alpha_l)w_l + (1 - \alpha_l) \sum_{k=x}^{N} w_k \cdot \mathbb{P}\big(p_k \in \widetilde{P_x} \mid p_k \in P_x\big) + \quad (6.2.9)$$
$$\alpha_l \sum_{k=l+1}^{N} w_k \cdot \mathbb{P}\big(p_k \in \widetilde{P_{l+1}} \mid p_k \in P_{l+1}\big).$$

And regrouping elements in (6.2.9) we obtain (6.2.4) for the case $p_l \in P_l$.

(ii) If $p_l \notin P_l$, from (6.1.6) we have:

$$\|P_l\|_{\mathbb{E}} = \sum_{k=l}^{N} w_k \cdot \mathbb{P}\big(p_k \in \widetilde{P_l} \mid p_k \in P_l, \, p_l \notin P_l\big). \quad (6.2.10)$$

Given that for this case $p_l \notin P_l$, we have that $p_l \notin \widetilde{P_l}$. Therefore:

$$\|P_l\|_{\mathbb{E}} = \sum_{k=l}^{N} w_k \cdot \mathbb{P}\big(p_k \in \widetilde{P_l} \mid p_k \in P_l, \, p_l \notin \widetilde{P_l}\big). \quad (6.2.11)$$

Substituting (6.2.8) in (6.2.11) yields (6.2.4) for $p_l \notin P_l$, completing the proof.

Finally, the computation of $\|P\|_{\mathbb{E}}$ requires the iterative calculation of the expected metric of the sub-schedules $P_N, P_{N-1}, \ldots, P_1$, which therefore takes $O(N)$. □

The algorithm presented in Lemma 6.2 for the calculation of the metric of a schedule can be modified to find the robust schedule in scenarios with a single scheduling entity:

Theorem 6.1. *The robust schedule for a set of passes P with a single scheduling entity can be computed recursively in $O(N)$ as $P^{\mathrm{R}} = P_1^{\mathrm{R}}$ where:*

$$P_l^{\mathrm{R}} = \begin{cases} \{p_l\} \cup P_{l+1}^{\mathrm{R}}, & \|\{p_l\} \cup P_{l+1}^{\mathrm{R}}\|_{\mathbb{E}} \geqslant \|P_{l+1}^{\mathrm{R}}\|_{\mathbb{E}}, \\ P_{l+1}^{\mathrm{R}}, & \text{otherwise}, \end{cases} \quad (6.2.12)$$

for all $p_l \in P$, and where P_l^{R} is the robust schedule for P_l.

Proof. We prove optimality by induction. From Lemma 6.2, we have:

$$\|\{p_l\} \cup P_{l+1}^{\mathrm{R}}\|_{\mathbb{E}} = (1 - \alpha_l)(w_l + \|D_l \cap P_{l+1}^{\mathrm{R}}\|_{\mathbb{E}}) + \alpha_l \|P_{l+1}^{\mathrm{R}}\|_{\mathbb{E}}, \quad (6.2.13)$$

and introducing (6.2.13) in (6.2.12):

$$\|P_l^{\mathrm{R}}\|_{\mathbb{E}} = \max \begin{cases} (1 - \alpha_l)(w_l + \|D_l \cap P_{l+1}^{\mathrm{R}}\|_{\mathbb{E}}) + \alpha_l \|P_{l+1}^{\mathrm{R}}\|_{\mathbb{E}}, \\ \|P_{l+1}^{\mathrm{R}}\|_{\mathbb{E}}. \end{cases} \quad (6.2.14)$$

Let $P^{\mathrm{R}} \subset P$ and $P_l^{\mathrm{R}} = P_l \cap P^{\mathrm{R}}$, and since $p_l \notin D_l$ we have that $D_l \cap P_l^{\mathrm{R}} = D_l \cap P_{l+1}^{\mathrm{R}}$. For a single scheduling entity we have from Proposition 6.1 that $D_l \cap P_{l+1}^{\mathrm{R}} = P_x^{\mathrm{R}}$, where p_x is the first pass non-conflicting with p_l that is in P_{l+1}^{R}. From (6.2.14) the expected metric of the schedule $\|P_l^{\mathrm{R}}\|_{\mathbb{E}}$ is maximal if $\|P_x^{\mathrm{R}}\|_{\mathbb{E}}$ and $\|P_{l+1}^{\mathrm{R}}\|_{\mathbb{E}}$ are maximal, which is equivalent to stating that P_l^{R} is robust for P_l if P_x^{R} and P_{l+1}^{R} are robust for P_x and P_{l+1}, respectively. Since p_N has no later conflicting passes, it is easy to see that $P_N^{\mathrm{R}} = \{p_N\}$ is robust for P_N. Therefore $P_1^{\mathrm{R}} = P^{\mathrm{R}}$ is robust.

The calculation of the robust schedule is iterative ($l = N, N-1, \ldots, 1$), and thus the calculation of P^{R} takes $O(N)$. The pass p_l is included in P_l^{R} if that increases the expected metric of the schedule, that is, if $\|\{p_l\} \cup P_{l+1}^{\mathrm{R}}\|_{\mathbb{E}} \geqslant \|P_{l+1}^{\mathrm{R}}\|_{\mathbb{E}}$.   □

Therefore, *the Robust SRS problem with a single ground station or satellite and a set of N passes $P = \{p_1, p_2, \ldots, p_N\}$ with associated weights w_i, failure probabilities $\alpha_i \in [0, 1]$ and fixed times (t_{s_i}, t_{e_i}) $\forall p_i \in P$ can be solved in $O(N)$.*

Let us now consider that the failure probabilities were not known. Then we could consider the worst case, which is $\alpha_l \to 1$ $\forall p_l \in P$. Under this assumption, if we calculate the robust schedule (Theorem 6.1), we have that the first line of (6.2.14) coincides with the second one: $\alpha_l = 1$ $\forall p_l \in P \Rightarrow (1-\alpha_l)(w_l + \|D_l \cap P_{l+1}^{\mathrm{R}}\|_{\mathbb{E}}) + \alpha_l \|P_{l+1}^{\mathrm{R}}\|_{\mathbb{E}} = \|P_{l+1}^{\mathrm{R}}\|_{\mathbb{E}}$, and thus $\|P_l^{\mathrm{R}}\|_{\mathbb{E}} = \|P_{l+1}^{\mathrm{R}}\|_{\mathbb{E}}$, so that we always add the pass under consideration to the schedule (6.2.12), that is: $\|P_l^{\mathrm{R}}\|_{\mathbb{E}} = \|P_{l+1}^{\mathrm{R}}\|_{\mathbb{E}} \Rightarrow P_l^{\mathrm{R}} = \{p_l\} \cup P_{l+1}^{\mathrm{R}}$ $\forall p_l \in P$. Therefore, $P^{\mathrm{R}} = P$ if $\alpha_l \to 1$ $\forall p_l \in P$.

6.3   Remarks regarding Multiple Scheduling Entities

We have seen in previous chapters that having a fixed number of scheduling entities would in some cases allow us to find tractable solutions. We show however that, even under this assumption, robust SRS is intractable for multiple entities.

Theorem 6.2. *The robust no-slack SRS problem is NP-hard for multiple ground stations and satellites ($|G| > 1$ and $|S| > 1$), even if these numbers are fixed.*

Proof. Finding the schedule with maximal expected metric has to be at least as complicated as calculating the metric of a single schedule. We will show that computing this metric for $|G| > 1$ and $|S| > 1$ is equivalent to probabilistic inference in multiply connected belief networks , which is NP-hard [4].

First, we generate the belief network. We create one node n_l for every pass p_l in the schedule P', each node representing a variable with two possible values: "true" and "false," $n_l = \{T, F\}$. For every pair of nodes n_i, n_j in that schedule, we place an edge from node n_i to node n_j if $\phi(p_j, p_i) = 1$ (so that the graph is directed); and from (6.2.2) only edges are created from passes with earlier start times to those with later start times and which are conflicting (so that the generated graph is also acyclic). The generation of this DAG takes $O(N^2)$. We assign to every node the probability:

$$\mathbb{P}\big(n_l = T\big) = (1 - \alpha_l) \cdot \mathbb{P}\big(n_m = F \ \forall n_m : \phi(p_m, p_l) = 1\big), \qquad (6.3.1)$$

which is equivalent to that from (6.1.5), so that $\mathbb{P}\big(n_l = T\big) = \mathbb{P}\big(p_l \in \widetilde{P'} \mid p_l \in P'\big)$. Thus we can represent the relations among the passes in a schedule into a belief network in polynomial time. From (6.1.6) the calculation of the metric $\|P'\|_{\mathbb{E}}$ requires calculating the conditional probabilities $\mathbb{P}\big(p_l \in \widetilde{P'} \mid p_l \in P'\big)$, which is equivalent to probabilistic inference in the generated belief network.

Second, we show that scenarios with multiple entities allow for loop cutsets (set of nodes which elimination would avoid the belief network to be multiply connected) to exist in the network (or equivalently in the DAG). Since we are assuming a fixed number (greater than one) of ground stations and satellites, we have that $|G| > 1$ and $|S| > 1$. Under this assumption, from (6.1.1) we have that time overlap is not a sufficient condition for them to be conflicting. Let us suppose there is such pair of passes (time overlapping but not conflicting). Then it would be possible for two passes that are not time overlapping among them, to be conflicting with the first pair. It is easy to see that the nodes associated to the second pair of passes would be connected by those associated to the first pair, and thus the network would be multiply connected.

Finally, Cooper [4] proves NP-hardness of probabilistic inference in multiply connected belief networks. Since we have shown that having multiple entities allows for the generated belief network to be multiply connected, and that the belief network can be generated in polynomial time, then for the worst case this computation will be NP-hard. Finding the optimal solution (robust schedule) has to be at least as complicated as calculating the expected metric of a given schedule, and therefore robust SRS is NP-hard for multiple ground stations and satellites. ☐

We show an example showing a multiple connected belief network generated from a set of passes for two satellites and two ground stations in Fig. 6.4.

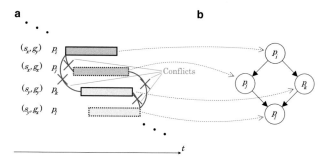

Fig. 6.4 Passes associated to a multiply connected belief network. (**a**) Set of passes. (**b**) Associated belief network

6.4   Variants of the Robust SRS Problem

In this section we present two variants of robust SRS: in the first case, we consider stochastic uncertainty in the priorities (§ 6.4.1); in the second case, we consider stochastic uncertainty in the pass duration (§ 6.4.2).

6.4.1   Robust SRS with Random Priorities

We modify the formulation presented in § 6.1 for considering the passes p_j^w, which instead of having associated failure probabilities, they have associated a set of possible priorities:

$$p_j^w = (s_i, g_h, t_{s_j}, t_{e_j}, W_j), \tag{6.4.1}$$

where W_j is a finite discrete random variable with the possible values for the priorities $w_{j,m}$ of the pass, each with probability $\beta_{j,m}$, that is:

$$W_j = \{w_{j,1} \text{ w.p. } \beta_{j,1}, \ w_{j,2} \text{ w.p. } \beta_{j,2}, \ \ldots, \ w_{j,M} \text{ w.p. } \beta_{j,M}\}, \tag{6.4.2}$$

such that $\sum_{m=1}^{M} \beta_{j,m} = 1$, and where $M = \max(|W_j|)$.

For every pass p_k with a smaller number of priorities $M_k < M$, we assume without loss of generality that $\beta_{k,m} = 0 \ \forall M_k < m \leqslant M$.

Definition 19. Let P^w be the set of passes with random priorities $\{p_j^w\}$, with: $p_j^w = (s_i, g_h, t_{s_j}, t_{e_j}, W_j)$, and where W_j is the set of priorities:

$$W_j = \{w_{j,1} \text{ w.p. } \beta_{j,1}, \ w_{j,2} \text{ w.p. } \beta_{j,2}, \ \ldots, \ w_{j,M} \text{ w.p. } \beta_{j,M}\}. \tag{6.4.3}$$

The problem of *robust SRS with random priorities* can be stated as finding the *robust schedule* P^{wR}, which is a schedule with maximal expected metric.

$$P^{wR} \subseteq P^w, \ \nexists P_{\text{sub}}^w \subseteq P^w : \|P_{\text{sub}}^w\|_\mathbb{E} > \|P^{wR}\|_\mathbb{E}. \tag{6.4.4}$$

In the remainder of this section we will tackle this problem through the framework presented for the problem with failure probabilities.

Proposition 6.2. *The robust SRS problem with failure probabilities is a subproblem of the robust SRS problem with random priorities.*

Proof. We will transform the set of passes with random priorities into precedence sets of passes with failure probabilities, but given that the passes in these groups will share the start and end times we will be able to provide an optimal solution in tractable time.

For each p_j^w we create a set P_j^w of M passes $p_{j,m}^w$ with the same start and end times as p_j^w, and with associated weights $w_{j,m}$ (6.4.2). These passes generated from the same p_j^w have to be considered together for inclusion in the final schedule, which can be modeled as precedence constraints (Definition 6, § 3.1.3). Let $P' \subset P_j^w$ be a schedule. We will calculate the failure probabilities of these passes $p_{j,m}^w$.

$$\beta_{j,m} = \mathbb{P}\big(p_{j,m} \in \widetilde{P'} \mid p_j^w \in P'\big) =$$
$$\mathbb{P}\big(p_{j,m} \in \widetilde{P'}, \, p_{j,m-1} \notin \widetilde{P'}, \, \ldots, \, p_{j,1} \notin \widetilde{P'} \mid p_j^w \in P'\big) =$$
$$\mathbb{P}\big(p_{j,m} \in \widetilde{P'} \mid p_{j,m-1} \notin \widetilde{P'}, \, \ldots, \, p_{j,1} \notin \widetilde{P'}, \, p_j^w \in P'\big) \cdot \qquad (6.4.5)$$
$$\mathbb{P}\big(p_{j,m-1} \notin \widetilde{P'}, \, \ldots, \, p_{j,1} \notin \widetilde{P'} \mid p_j^w \in P'\big).$$

The first term of the product (6.4.5) is the probability of that pass $p_{j,m}$ to be successful given that all the previous passes $p_{j,n} : n < m$ fail, that is:

$$\mathbb{P}\big(p_{j,m} \in \widetilde{P'} \mid p_{j,m-1} \notin \widetilde{P'}, \, \ldots, \, p_{j,1} \notin \widetilde{P'}, \, p_j^w \in P'\big) = 1 - \alpha_{j,m}. \qquad (6.4.6)$$

The second term of the product (6.4.5) is the probability of all the previous passes $p_{j,n} : n < m$ to fail, that is:

$$\mathbb{P}\big(p_{j,m-1} \notin \widetilde{P'}, \, \ldots, \, p_{j,1} \notin \widetilde{P'} \mid p_j^w \in P'\big) = \alpha_{j,m-1} \cdot \alpha_{j,m-2} \cdots \alpha_{j,1}. \qquad (6.4.7)$$

Therefore:

$$\beta_{j,m} = (1 - \alpha_{j,m}) \prod_{k=1}^{m-1} \alpha_{j,k}, \qquad (6.4.8)$$

so that we can calculate the values of $\alpha_{j,m}$ recursively:

$$\alpha_{j,m} = 1 - \beta_{j,m} \prod_{k=1}^{m-1} \alpha_{j,k}^{-1}. \qquad (6.4.9)$$

Note that these passes generated from the same p_j^w will be mutually conflicting. They also share start and end times, so we could add small time increments to their start and end times to keep the ordering. For the sake of simplicity we will just assume that the ordering is kept based on the index of the generated passes, so that:

$$\phi(p_{j,n}^w, p_{j,m}^w) = 1, \quad \phi(p_{j,m}^w, p_{j,n}^w) = 0 \Leftrightarrow m < n. \qquad (6.4.10)$$

And thus we define the sets P_j^w of passes associated to the same random-priority pass p_j^w as follows:

$$\forall p_j^w \in P^w : p_j^w = (s_i, g_h, t_{s_j}, t_{e_j}, W_j),$$
$$\exists P_j^w = \{p_{(j-1)\cdot M+m}\} : m = 1, 2, \ldots, M, \qquad (6.4.11)$$
$$p_{(j-1)\cdot M+m} = p_{j,m}^w = (s_i, g_h, t_{s_j}, t_{e_j}, w_{j,m}, \alpha_{j,m}),$$

where the values for $w_{j,m}$ are obtained from (6.4.2), and those for $\alpha_{j,m}$ from (6.4.9). Then the final set of passes is:

$$P = \bigcup_{j=1}^{N} P_j^{\text{w}} \text{ with precedence subsets } P_j^{\text{w}}. \tag{6.4.12}$$

Therefore robust SRS with random priorities can be transformed into robust SRS with failure probabilities, but with precedence constraints. From Definition 6 (§ 3.1.3) no-precedence is a subproblem of precedence, and therefore robust SRS with failure probabilities is a subproblem of robust SRS with random priorities. It is easy to see that generating each of the sets P_j^{w} takes $O(M)$, and since there are N random-priority passes, the transformation takes $O(NM)$. □

The presented transformation allows to solve this variant of robust SRS with an extension of the algorithm in Theorem 6.1. For a summary of this extension see Fig. 6.6.

Corollary 6.2. *The robust SRS problem with random priorities for a single scheduling entity and a set of passes $P^{\text{w}} = \{p_1^{\text{w}}, p_2^{\text{w}}, \ldots, p_N^{\text{w}}\}$ with random weights $W_i = \{w_{i,1} \text{ w.p. } \beta_{i,1}, w_{i,2} \text{ w.p. } \beta_{i,2}, \cdots, w_{i,M} \text{ w.p. } \beta_{i,M}\}$ and fixed times $(t_{s_i}, t_{e_i}) \forall p_i^{\text{w}} \in P^{\text{w}}$ can be solved in $O(NM)$, where N is the number of passes, and M is the maximum number of weights for W_i.*

Proof. Let $|P^{\text{w}}| = N$, and $|W_i| = M$. We generate the sets $P_j^{\text{w}} \forall p_j^{\text{w}} \in P^{\text{w}}$ as in (6.4.11), so that we obtain the set of passes P through (6.4.12).

Since every set P_j^{w} is a precedence subset, either all its nodes are included in the schedule or none, so that we only have to modify the algorithm presented in Theorem 6.1 for updating the robust schedule only when evaluating the first passes of the sets, that is $p_{(j-1)\cdot M+1} \forall j = 1, 2, \ldots, N$, or equivalently for those passes $p_l \in P$ for which $l - 1$ is divisible by M:

$$P_l^{\text{R}} = \begin{cases} \begin{cases} \{p_l\} \cup P_{l+1}^{\text{R}}, & \text{if } \|\{p_l\} \cup P_{l+1}^{\text{R}}\|_{\text{E}} \geq \|P_{l+M}^{\text{R}}\|_{\text{E}}, \\ P_{l+M}^{\text{R}}, & \text{otherwise,} \end{cases} & \text{if } M \mid l-1, \\ \{p_l\} \cup P_{l+1}^{\text{R}}, & \text{otherwise,} \end{cases} \tag{6.4.13}$$

for all $p_l \in P$, with $\|P_l^{\text{R}}\|_{\text{E}} = 0 \ \forall p_l \notin P$ for consistency of the algorithm, and where P_l^{R} is the robust schedule for P_l. As for Theorem 6.1, introducing (6.2.13) in (6.4.13):

$$\|P_l^{\text{R}}\|_{\text{E}} = \begin{cases} \max \begin{cases} (1 - \alpha_l)(w_l + \|D_l \cap P_{l+1}^{\text{R}}\|_{\text{E}}) + \alpha_l \|P_{l+1}^{\text{R}}\|_{\text{E}}, \\ \|P_{l+M}^{\text{R}}\|_{\text{E}}, \end{cases} & \text{if } M \mid l-1, \\ (1 - \alpha_l)(w_l + \|D_l \cap P_{l+1}^{\text{R}}\|_{\text{E}}) + \alpha_l \|P_{l+1}^{\text{R}}\|_{\text{E}}, & \text{otherwise.} \end{cases} \tag{6.4.14}$$

The new expressions (6.4.13) and (6.4.14) guarantee that the sets P_j^w are either completely included or dismissed in the robust schedule:

- If the evaluated pass is not the first of a subset ($M \nmid l-1$), then the pass is added to the robust schedule and the metric recalculated.
- But if the evaluated pass is the first of a subset ($M \mid l-1$), then the pass is added to the schedule if the new metric is higher than that associated to the first pass of the next set, otherwise the metric before adding the last pass (first added) of the evaluated precedence set is taken again.

Proofs for correctness and complexity are equivalent to those from Theorem 6.1. Since $|P| = NM$, the robust schedule for the robust SRS problem with random priorities for a single scheduling entity can be solved in $O(NM)$.   □

6.4.2   Robust SRS with Random Durations

In this case we consider an initial set of requests with random durations, so that the scheduler may specify the times when passes will start, but their duration will be uncertain. The approach we will follow is to discretize time and generate the equivalent passes with failure probabilities, similarly as in the previous section. Let this set be P^d. Again passes generated from the same request will be in the same precedence set, but in this case the intervals will have in general different start and end times. We have shown that problems with precedence are intractable in general, and in the remainder of the section we will prove intractability for this case.

Let J^w be the initial set of requests j_j^w for the problem, given a time step Δt, such that the possible durations for the pass are consecutive multiples of this step $\rho_{j,k} = n_{j,k} \Delta t \ \forall k = 1, 2, \ldots, M$, and with $J^w = \{j_j^w\}$. Let the request j_j^w be:

$$j_j^w \triangleq (s_h, g_i, r_j, d_j, \Gamma_j, W_j) :$$
$$\Gamma_j = \{\rho_{j,1}, \rho_{j,2}, \ldots, \rho_{j,M}\} : \rho_{j,1} \leqslant \rho_{j,2} \leqslant \cdots \leqslant \rho_{j,M}, \qquad (6.4.15)$$
$$W_j = \{w_{j,1} \text{ w.p. } \beta_{j,1}', \ w_{j,2} \text{ w.p. } \beta_{j,2}', \ \ldots, \ w_{j,M} \text{ w.p. } \beta_{j,M}', \},$$

such that $\sum_{m=1}^{M} \beta_{j,m}' = 1$, with $n_{j,k} \in \mathbb{N}$, and for the worst case $M = \max(\lfloor \frac{d_j - r_j}{\Delta t} \rfloor)$.
Also, if Γ_j has $M' < M$ possible values, then $\beta_{j,m}' = 0 \ \forall M' < m \leqslant M$.

Definition 20. The *robust SRS problem with random durations* can be stated as finding the *robust schedule* P^{dR}, which is a schedule with maximal expected metric.

$$P^{dR} \subseteq P^d, \ \nexists P_{sub}^d \subseteq P^d : \|P_{sub}^d\|_{\mathbb{E}} > \|P^{dR}\|_{\mathbb{E}}. \qquad (6.4.16)$$

We now describe the transformation of the set of requests J^w into the set of passes with random probabilities P^d.

Proposition 6.3. *The robust SRS problem with failure probabilities is a subproblem of the robust SRS problem with random durations.*

Proof. The proof follows the same idea as that for Proposition 6.2. We generate the extended set of passes through Transformation 1 from Chap. 3, which creates a set of passes with a duration multiple of the considered discretization step Δt complying with the constraints on the duration of the requests:

$$D_n(J^{\mathrm{w}}) = P^{\mathrm{d}}. \tag{6.4.17}$$

We regroup all the generated passes into the sets $P^{\mathrm{w}}_{j,k}$ such that two passes belong to the same set if they originate from the same request and have the same start time:

$$p_{j,k,m}, p_{j,k,n} \in P^{\mathrm{w}}_{j,k} \Leftrightarrow \{D_n^{-1}(p_{j,k,m}) = D_n^{-1}(p_{j,k,n})\} \wedge \{t_{\mathrm{s}}(p_{j,k,m}) = t_{\mathrm{s}}(p_{j,k,n})\}. \tag{6.4.18}$$

Since $|\Gamma_j| = M$, all the start and end times $\forall p_{j,k,m} \in P^{\mathrm{w}}_{j,k}$ follow the form:

$$t_{\mathrm{s}}(p_{j,k,m}) = r_j + (k-1)\Delta t, \tag{6.4.19}$$

$$t_{\mathrm{e}}(p_{j,k,m}) = r_j + (k-1+m)\Delta t, \tag{6.4.20}$$

with $m \leq (d_j - r_j)\Delta t^{-1} - k + 1$ and therefore $|P^{\mathrm{w}}_{j,k}| = M - k + 1$. We assume that the passes in $P^{\mathrm{w}}_{j,k}$ are ordered by increasing end time, with the same considerations for conflicts as in (6.4.10): $\forall p_{j,k,m}, p_{j,k,n} \in P^{\mathrm{w}}_{j,k}, \ m < n \Leftrightarrow t_{\mathrm{e}}(p_{j,k,m}) < t_{\mathrm{e}}(p_{j,k,n})$.

We now calculate the failure probabilities of these passes similarly as in the previous section. These subsets $P^{\mathrm{w}}_{j,k}$ do not have all the possible durations but only those from Δt to $(M - k + 1)\Delta t$, so that we have to adjust the probabilities of their passes. Let $P^{\mathrm{w}}_{j,k} = \{p_{j,k,1}, p_{j,k,2}, \ldots, p_{j,k,M-k+1}\}$. For each of the passes $p_{j,k,m}$ in the set $P^{\mathrm{w}}_{j,k}$ we assign the modified priorities $\beta'_{j,k,m}$:

$$\beta'_{j,k,m} = \beta'_{j,m}\Big(\sum_{n=1}^{M-k+1} \beta'_{j,n} \Big)^{-1}, \tag{6.4.21}$$

so that $\sum_{m=1}^{M-k+1} \beta'_{j,k,m} = 1$. Therefore, from (6.4.9):

$$\alpha_{j,k,m} = 1 - \beta'_{j,k,m} \prod_{x=1}^{m-1} \alpha_{j,k,x}^{-1}. \tag{6.4.22}$$

The priorities only depend on the duration of the request (6.4.15), so we have that:

$$w_{j,k,m} = w_{j,m}. \tag{6.4.23}$$

Note that since the durations of these passes are random, the sets $P^w_{j,k}$ must be considered completely either for inclusion or omission in the robust schedule. Then the final set of passes is:

$$P^d = \bigcup_{j=1}^{N} \bigcup_{k=1}^{M} P^w_{j,k} \text{ with precedence constraints } P^w_{j,k}, \qquad (6.4.24)$$

and with every pair of passes in $\bigcup_{k=1}^{M} P_{j,k}$ being considered conflicting even if they are not time overlapping, because they are originated from the same request. Given that $|J^w| = N$ and $|\Gamma_j| = M \; \forall j$, from (3.1.14) this transformation takes $O(NM^2)$. □

As we will show, this problem also generalizes that with random priorities.

Proposition 6.4. *The robust SRS problem with random priorities is a subproblem of the robust SRS problem with random durations.*

Proof. It is trivial to see that if we constrain the possible values of the durations for each request to be all equal and maximum, the robust SRS problem with failure probabilities corresponds to the robust SRS problem with random priorities. Each pass $p^w_j = (s_i, g_h, t_{s_j}, t_{e_j}, W_j)$ can be modeled as a request $j^w = (s_i, g_h, r_j, d_j, \Gamma_j, W_j)$ where $r_j = t_{s_j}$, $d_j = t_{e_j}$, and all the values of the elements in Γ_j are equal to $t_{e_j} - t_{s_j}$. □

Note that the transformation in Proposition 6.3 generates passes which are conflicting although they are not time overlapping, and therefore the condition stated in Proposition 6.1, which is the base for the algorithms presented in § 6.2 and § 6.4.1 is not applicable. We now show that robust SRS with random durations is intractable.

Corollary 6.3. *The robust SRS problem with random durations for a single scheduling entity and a set of requests J^w with random durations $\Gamma_i = \{\rho_{i,1}, \rho_{i,2}, \ldots, \rho_{i,M}\}$ and associated weights $W_i = \{w_{i,1} \text{ w.p. } \beta_{i,1}, w_{i,2} \text{ w.p. } \beta_{i,2}, \cdots, w_{i,M} \text{ w.p. } \beta_{i,M}\}$ is NP-hard.*

Proof. This proof is similar to that of Theorem 6.2. We will show NP-hardness for a subproblem of robust SRS with random durations by showing that the belief network generated by the initial set of requests may be multiply connected.

Let us consider the subproblem of robust SRS with random durations where passes (generated from the requests) can either have a fixed duration smaller than the duration of the request or may fail (zero duration and zero priority), that is:

$$j^w_j \triangleq (s_h, g_i, r_j, d_j, \Gamma_j, W_j), \qquad (6.4.25)$$

with the pair (Γ_j, W_j) taking two possible values:

$$(\Gamma_j, W_j) = \begin{cases} (0,0) & \text{w.p. } \alpha_j, \\ (\rho_j, w_j) & \text{w.p. } 1 - \alpha_j, \end{cases} \qquad (6.4.26)$$

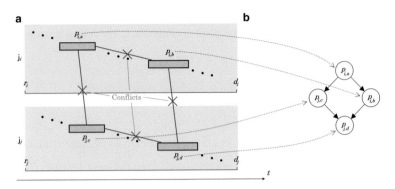

Fig. 6.5 Passes generated from two identical requests which are associated to a multiply connected belief network. (**a**) Set of passes. (**b**) Associated belief network

and we force the generation of passes conflicting even if not time overlapping for the same request, by selecting a duration ρ_j such that $2 \cdot \rho_j < d_j - r_j$.

Assuming that $|G| = 1$ and $|S| > 1$, there could be time overlapping requests in J^w. Let us then consider a pair of requests overlapping in time, each generating at least a pair of passes (note the constraint on ρ_j). If the earliest start time passes of the two pairs are mutually conflicting, and so are the latest start time ones, but the earliest and latest passes of different pairs do not conflict, the belief network that these passes generate would be multiply connected. This scenario is easily achieved with duplicated requests.

Finally, Cooper [4] proves NP-hardness of the calculation of these probabilities in multiply connected belief networks. Finding the robust schedule has to be at least as complicated as calculating its expected metric, which from Definition 16 requires calculating these probabilities, and thus for the worst case the problem is NP-hard.

□

We show an example showing a multiple connected belief network generated from two identical requests in Fig. 6.5.

The condition stated in Proposition 6.1 would be however applicable to the random durations problem if preemption were allowed. In this case the set of passes generated through Proposition 6.3 would be the same as that in (6.4.24) but removing additional considerations on the conflicts among passes: $P'^{d} = \bigcup_{j=1}^{N} \bigcup_{k=1}^{M} P_{j,k}^w$ with precedence sets $P_{j,k}^w$. Since we allow for preemption, the passes in P'^d comply with the property described in Proposition 6.1, and therefore we can apply the algorithm in Corollary 6.2. Since from Proposition 6.3 we have that $|P'^d| = O(NM^2)$, then it is easy to see that applying the algorithm to P'^d takes $O(NM^2)$.

6.5 Considerations on the Basic SRS Problem

In this section we provide an algorithm for the single scheduling entity basic SRS problem with priorities $1 \mid \tau_j, \overline{\overline{\mathfrak{p}_{ij}}} \mid \sum \mathfrak{w}_j \mathfrak{U}_j$ (§ 3.3), based on the algorithm in § 6.2.

Theorem 6.3. *The SRS problem with a single ground station or satellite and a set of passes* $P = \{p_1, p_2, \ldots, p_N\}$ *with associated weights* w_i *and fixed times* (t_{s_i}, t_{e_i}) $\forall p_i \in P$ *can be solved in* $O(N)$, *where* N *is the number of passes.*

Proof. From Lemma 6.1, this problem is a subproblem of the analog version in robust SRS with failure probabilities, where all these probabilities are null. Therefore, we can apply the algorithm in Theorem 6.1 for finding the robust schedule P^R in $O(N)$.

As seen previously this robust schedule is not necessarily feasible, but from Proposition 6.1 we can examine the set P^R by increasing start time: for each examined pass, we delete all the subsequent passes conflicting with it, and keep examining the set after the last removed pass; the remainder of P^R after this computation will be the optimal schedule.

From Proposition 6.1 once we find a non-conflicting pass while examining P^R, the rest of the passes will not be conflicting either, so that this computation takes $O(N)$.   □

Figure 6.6 represents the relations between the algorithms presented in this chapter for robust SRS.

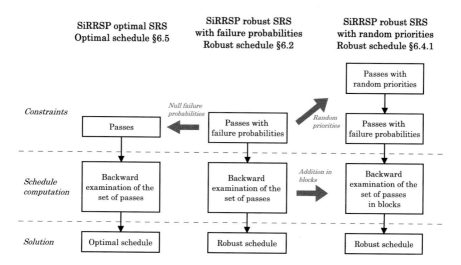

Fig. 6.6 Relations between the algorithms for robust SRS

6.6 Schedule Computation Example

This example aims to explain the proposed algorithm for the calculation of the robust schedule in a simple scenario with a single ground station and where passes have failure probabilities. The pass intervals are represented in Fig. 6.7, where passes are ordered by their start times. For simplicity we consider that the failure probabilities are the same for all the passes: $\alpha_l = 0.5 \; \forall p_l \in P$.

We follow the algorithm in Theorem 6.1 for the calculation of the robust schedule, so that we will evaluate the passes for inclusion in the optimal schedule starting from p_6. For this pass, we have that $P_6^{\mathrm{R}} = \{p_6\}$ and $\|P_6^{\mathrm{R}}\|_{\mathbb{E}} = (1-\alpha_6)w_6 = 0.35$.

We then evaluate pass p_5. If we added it to the schedule then the expected metric would be $(1-\alpha_5)w_5 + \alpha_5\|P_6^{\mathrm{R}}\|_{\mathbb{E}} = 0.18$, which is less than $\|P_5^{\mathrm{R}}\|_{\mathbb{E}} = 0.35$ with $P_5^{\mathrm{R}} = \{p_6\}$, so that the pass is not included in the schedule.

The next pass to consider is p_4, which is added to the schedule since it has not any later pass conflicting with it. Therefore $P_4^{\mathrm{R}} = \{p_4, p_6\}$, and its associated metric $\|P_4^{\mathrm{R}}\|_{\mathbb{E}} = (1-\alpha_4)w_4 + \|P_5^{\mathrm{R}}\|_{\mathbb{E}} = 0.55$.

The pass p_3 is added to the schedule following the same reasoning as for p_4, so that $P_3^{\mathrm{R}} = \{p_3, p_4, p_6\}$ and $\|P_3^{\mathrm{R}}\|_{\mathbb{E}} = (1-\alpha_3)w_3 + \|P_4^{\mathrm{R}}\|_{\mathbb{E}} = 0.8$.

The pass p_2 will also be included in the schedule since $\|P_2^{\mathrm{R}}\|_{\mathbb{E}} = 1.025$ with $P_2^{\mathrm{R}} = \{p_2, p_3, p_4, p_6\}$. Since the first pass in $D_2 \cap P_3^{\mathrm{R}}$ is p_6, this new metric can be calculated as $(1-\alpha_2)(w_2 + \|P_6^{\mathrm{R}}\|_{\mathbb{E}}) + \alpha_2\|P_3^{\mathrm{R}}\|_{\mathbb{E}}$.

Finally we dismiss the pass p_1, so that $P_1^{\mathrm{R}} = \{p_2, p_3, p_4, p_6\}$ and therefore $\|P_1^{\mathrm{R}}\|_{\mathbb{E}} = 1.025$. Otherwise, since the first pass in $D_1 \cap P_2^{\mathrm{R}}$ is p_4, the expected metric would be $(1-\alpha_1)(w_1 + \|P_4^{\mathrm{R}}\|_{\mathbb{E}}) + \alpha_1\|P_2^{\mathrm{R}}\|_{\mathbb{E}} = 0.8875$, which is less than $\|P_1^{\mathrm{R}}\|_{\mathbb{E}}$.

Thus $P^{\mathrm{R}} = \{p_2, p_3, p_4, p_6\}$ and $\|P^{\mathrm{R}}\|_{\mathbb{E}} = 1.025$. The optimal algorithm yields $P^* = \{p_2, p_6\}$, with $\|P^*\|_{\Sigma w} = 1.6$ and $\|P^*\|_{\mathbb{E}} = 0.8$, and the greedy earliest start time algorithm yields the metric $\|P\|_{\mathbb{E}} = 0.67$.

6.7 Simulations

In this section we provide simulations for showing the performance of the algorithm presented for robust SRS with failure probabilities in practical scenarios. We include the results of another two reference algorithms in SRS: *optimal* algorithm for basic

Fig. 6.7 Set of passes for a single ground station. [3] ©2015 IEEE

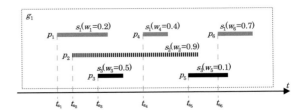

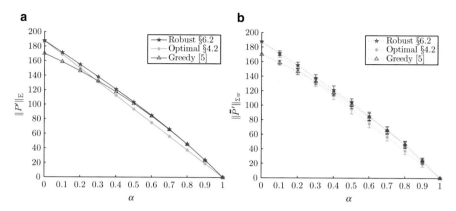

Fig. 6.8 Simulation results for scenario 1. (**a**) Expected metrics. (**b**) Metrics of executed schedules. [3] ©2015 IEEE

SRS (§ 4.2) (for which the selected schedule is P^*), and the *greedy earliest start time* algorithm (see, for example, [5], for which the selected schedule is P).

We implemented the three algorithms (optimal, greedy, and robust) in MATLAB to evaluate their performance with different values for the failure probabilities. We consider the two scenarios we introduced in § 4.6 but considering a single ground station and a fixed scheduling horizon of 14 days. Both scenarios have $|G| = 1$, $|S| = 5$, a scheduling horizon of 14 days, fixed interval passes with random priorities $w(p_l) = v_l/10 : v_l \sim U[1,10]$, and we assign the same failure probability to all the passes ($\alpha_l = \alpha \ \forall l$) varying α between 0 and 1 in steps of 10^{-1}. Whereas for scenario 1 we consider satellites in different low Earth orbits (LEO), scenario 2 corresponds to the worst case scenario for the presented algorithm, wherein all the orbits are identical; so that this case yields the maximum number of conflicts.

For the implementation we calculate the expected metric of the three schedules $\|P^*\|_{\mathbb{E}}$ (optimal), $\|P\|_{\mathbb{E}}$ (greedy), and $\|P^R\|_{\mathbb{E}}$ (robust) through the algorithm in Lemma 6.2. We show the results for scenarios 1 and 2 in Figs. 6.8a and 6.9a, respectively. Alternatively, we simulate 100 realizations of the metric of the executed schedules $\|\widetilde{P^*}\|_{\Sigma w}$, $\|\tilde{P}\|_{\Sigma w}$ and $\|\widetilde{P^R}\|_{\Sigma w}$ for the same range of values of α, based on (6.1.5) and (6.1.3). We show the confidence intervals of these results in Figs. 6.8b and 6.9b.

The obtained results show that $\|P^*\|_{\mathbb{E}} \leq \|P^R\|_{\mathbb{E}}$ and $\|P\|_{\mathbb{E}} \leq \|P^R\|_{\mathbb{E}}$, with the biggest differences in metric in the scenario where positions of the ground stations and orbits of the satellites are respectively highly correlated.

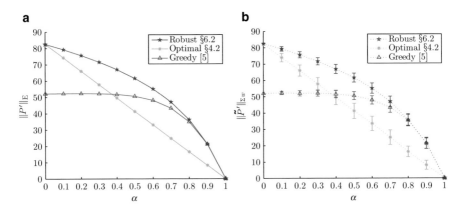

Fig. 6.9 Simulation results for scenario 2. (**a**) Expected metrics. (**b**) Metrics of executed schedules. [3] ©2015 IEEE

6.8   Summary

We extend the notation in Chap. 3 (§ 3.3) to cover the presented cases:

- *Failure probabilities*: we add the wide tilde symbol to the term in the optimization function indicating if the pass is in the schedule, that is, $\sum \mathfrak{w}_j \widetilde{\mathfrak{U}_j}$.
- *Random priorities*: similarly to the previous case we add the wide tilde symbol to the priorities term in the objective function, so that we have $\sum \widetilde{\mathfrak{w}_j} \mathfrak{U}_j$.
- *Random durations*: in this case it is the constraints term the one that receives the wide tilde: $\underline{\mathfrak{p}_{ij}} \leqslant \widetilde{\mathfrak{p}_{ij}} \leqslant \overline{\mathfrak{p}_{ij}}$.

We summarize the results presented in this chapter into Table 6.1 for easy consultation, including the references for some of the approximate solutions available in literature, and where N is the number of passes or requests and M is the fixed number of values for the priorities or the durations (which for this case is equivalent to the maximum number of time steps in a pass). We add the problem $\mathfrak{R}|\mathfrak{r}_j, \underline{\mathfrak{p}_{ij}} \leqslant \widetilde{\mathfrak{p}_{ij}} \leqslant \overline{\mathfrak{p}_{ij}}, C_{\Sigma}|\sum \mathfrak{w}_j \mathfrak{U}_j$ for completeness. It is easy to see that this problem generalizes $1|\mathfrak{r}_j, \underline{\mathfrak{p}_{ij}} \leqslant \widetilde{\mathfrak{p}_{ij}} \leqslant \overline{\mathfrak{p}_{ij}}|\sum \mathfrak{w}_j \mathfrak{U}_j$ and $\mathfrak{R}|\mathfrak{r}_j, \overline{\overline{\mathfrak{p}_{ij}}}, C_{\Sigma}|\sum \mathfrak{w}_j \widetilde{\mathfrak{U}_j}$, and therefore it is NP-hard.

The complexity of the presented variable slack problems is displayed for discrete time (Δt) and a fixed number of priorities or durations (M) as applicable. We show the relations among the presented problems in Fig. 6.10.

Acknowledgements This research was performed while the author held a National Research Council Research Associateship Award at the Air Force Research Laboratory (AFRL).

Table 6.1 Optimal solutions for Robust SRS

Problem	Resources		Slack		Uncertainty	Complexity $\Delta t, M$	References	
	Single	Multiple	No	Variable			Approx.	Optimal
$1\,\lvert r_j,\overline{\overline{p_{ij}}}\rvert\sum w_j \widetilde{\Delta t_j}$	x		x		Failure	$O(N)$	–	§ 6.2
$1\,\lvert r_j,\overline{\overline{p_{ij}}}\rvert\sum w_j \widetilde{\Delta t_j}$	x		x		Priority	$O(NM)$	–	§ 6.4.1
$1\,\lvert r_j,\underline{p_{ij}}\leqq p_{ij}\rvert\sum w_j \widetilde{\Delta t_j}$	x			x	Duration	NP-hard	[2]	§ 6.4.2
$\Re\,\lvert r_j,\overline{\overline{p_{ij}}},C_\Sigma\rvert\sum w_j \Delta t_j$		x	x		Failure	NP-hard	–	§ 6.3
$\Re\,\lvert r_j,\underline{p_{ij}}\leqq p_{ij},C_\Sigma\rvert\sum w_j \Delta t_j$		x		x	Duration	NP-hard	–	§ 6.8

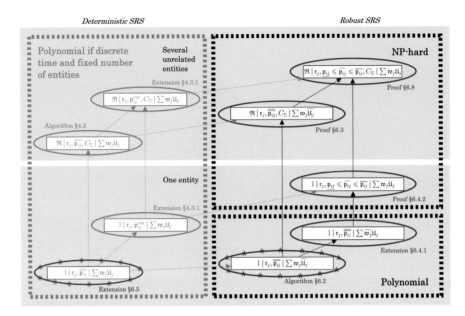

Fig. 6.10 Complexity of robust SRS and relations with deterministic SRS

References

1. S. Badaloni, M. Falda, M. Giacomin, Solving temporal over-constrained problems using fuzzy techniques. J. Intell. Fuzzy Syst. **18**(3), 255–265 (2007)
2. M. Drummond, J. Bresina, K. Swanson, Just-in-case scheduling, in *AAAI-94 Proceedings* (AAAI, Seattle, 1994)
3. A.J. Vazquez, R.S. Erwin, Robust fixed interval satellite range scheduling, in *2015 IEEE Aerospace Conference* (IEEE, Big Sky, 2015)
4. G.F. Cooper, The computational complexity of probabilistic inference using Bayesian belief networks. Artif. Intell. **42**, 393–405 (1990)
5. A. Globus, J. Crawford, J. Lohn, A. Pryor, A comparison of techniques for scheduling Earth observing satellites, in *Proceedings of the Sixteenth Innovative Applications of Artificial Intelligence Conference* (IAAI, San Jose, 2004)

Chapter 7
Reactive Satellite Range Scheduling

The last variant of the SRS problem treated in this book is the reactive SRS problem. This problem is closely related to the one presented in Chap. 6 (§ 6.4.1) where the priorities associated to the passes were random. The difference in this case is that these priorities are known at all times, but they dynamically change along the execution of the schedule. We assume that these changes cannot be predicted, and therefore the solution will be the reactive calculation of the schedule.

This problem has also been presented in previous SRS literature [1]. The satellite operators may need to raise the priority of a pass that becomes critical along the execution of the schedule, or a satellite or ground station may cancel a set of passes. Reactive scheduling would allow to adapt to these changes and avoid a reduction in performance, by speeding up the recomputation of the optimal schedule.

In this chapter we focus on speeding up this recomputation, and specifically we present a modification to the algorithm presented in Chap. 4 (§ 4.2) introducing a preprocessing phase for reducing the computation time for the recalculation of the optimal schedule after the passes change their priorities.

7.1 Scenario Model for Reactive SRS

The model for this problem is the same as that presented in Chap. 4 for the basic SRS problem, except that in this case the priorities of the passes vary along the execution of the schedule. As for this previous model we assume that all the entities will deterministically follow the same schedule, as shown in Fig. 7.1. The dynamic changes of the passes, which are known to the scheduler when they occur, are represented with dotted lines connecting the satellites with the scheduler.

Let $S = \{s_i\}$ be a set of satellites, and $G = \{g_h\}$ a set of ground stations. As for previous chapters, we consider a scheduling horizon T starting at t_0: $t \in [t_0, t_0 + T]$.

© Springer International Publishing Switzerland 2015
A.J. Vázquez Álvarez, R.S. Erwin, *An Introduction to Optimal Satellite Range Scheduling*, Springer Optimization and Its Applications 106,
DOI 10.1007/978-3-319-25409-8_7

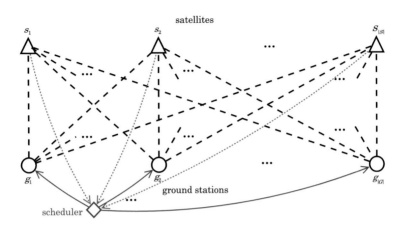

Fig. 7.1   Reactive scheduler

We consider that a pass may change its probability along time, so that:

$$p_l(t) = (s_i, g_h, t_{s_l}, t_{e_l}, w_l(t)). \tag{7.1.1}$$

Since the passes are time dependent, the initial set of passes is now $P(t)$.

Remark 7.1. We assume that only future passes can change priorities, so that we have that $w_l(t) = w_l(t_s(p_l)) \; \forall t \geq t_s(p_l), \; \forall p_l \in P(t)$.

Let $P(t)\big|_{t_1}^{t_2}$ be the subset of passes of $P(t)$ with start times in the interval $[t_1, t_2]$:

$$p_l \in P(t)\big|_{t_1}^{t_2} \; \Leftrightarrow \; p_l \in P(t) \; \wedge \; t_1 \leq t_s(p_l) \leq t_2. \tag{7.1.2}$$

Remark 7.2. In reactive SRS we recompute the optimal schedule with every priority change. That is, for discretized time with a time step Δt, there will be re-scheduling if $P(t) \neq P(t - \Delta t)$. Let τ_c be the minimum time between two changes in $P(t)$. We assume that $\Delta t < \tau_c$, otherwise we could select a smaller discretization step.

Definition 21. The *reactive SRS* problem can be stated as finding the *updated optimal schedule $P^*(t)$ minimizing the computation time*, which is a schedule with maximal metric and keeping the part of the schedule already executed.

$$P^*(t) \triangleq P^*(t - \Delta t))\big|_{t_0}^{t - \Delta t} \cup P^*(t)\big|_{t}^{t_0 + T} \; : \; P^*(t) \in \{P^f\}. \tag{7.1.3}$$

In basic SRS the schedule is calculated once before its execution, and the obtained performance is the same as the expected. If we introduce changes in priority but the schedule is not updated, then the performance may be different than expected, and there will likely be a new optimal schedule different from the current one. From Remark 7.1 we have that $P^*(t)\big|_{t_0}^{t - \Delta t} = P^*(t - \Delta t)\big|_{t_0}^{t - \Delta t}$, so that

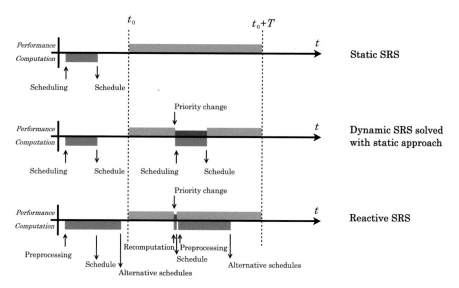

Fig. 7.2 Examples of scenarios for static and dynamic SRS: static SRS (*top*), dynamic SRS solved with static approach (*center*), and reactive SRS (*bottom*); indicating periods where computations are required for finding new/alternative schedules (*blue*), periods where the schedule is executing optimally with respect to the current set of priorities (*green*), and periods where the schedule is executing suboptimally after a change in priorities (*red*)

the reference solution for this problem would be applying the optimal algorithm (§ 4.2) to the set $P(t)|_t^{t_0+T}$. However, this would take $O(N(k_1 + 1)^{k_2})$ for the worst case. In practical problems only a small number of passes change priorities at the same time, fact which raises the question: is it possible to reduce this computation time?

In reactive scheduling we aim at introducing a *preprocessing* stage which allows to speed up the *recomputation* of the optimal schedule with the new priorities. We illustrate this problem in Fig. 7.2.

Preprocessing	The preprocessing includes the calculations performed prior to the priority change in order to reduce the recalculation time from the time the change is known.
Recomputation	The recomputation includes those calculations needed to find the new optimal schedule once the priority change is known.

7.1.1   Complexity of the Reactive SRS Problem

In this section we study the relation of the reactive SRS problem with some of the problems presented in previous chapters.

Lemma 7.1. *The reactive SRS problem generalizes the basic SRS problem.*

Proof. We have presented in Remark 7.2 a constraint on the time between changes (τ_c). It is easy to see that if $\tau_c \to \infty$, then $P(t)$ is constant, and therefore the problem reduces to the basic SRS problem $\mathfrak{R} \mid r_j, \overline{\overline{p_{ij}}}, C_\Sigma \mid \sum w_j \mathfrak{U}_j$ (§ 3.3). □

The problem would also reduce to basic SRS if the priorities $w_l(t)$ where known a priori, as this would allow for the computation of the priorities $w_l = w_l(t_{s_l})$, which would therefore be static. Reactive SRS is also closely related to robust SRS:

Lemma 7.2. *The reactive SRS problem generalizes the robust SRS problem with random priorities.*

Proof. In this case, if $\tau_c \to 0$, then the priority of a pass can be considered unknown until the pass starts. If the model for the changes in the priorities were available, then the problem would reduce to the robust SRS problem with random priorities $\mathfrak{R} \mid r_j, \overline{\overline{p_{ij}}} \mid \sum \widetilde{w}_j \mathfrak{U}_j$; and if no model were available, then we would follow the same reasoning as in the last paragraph of § 6.2. □

Some references [2] consider reactive and robust scheduling to be both types of dynamic scheduling. In this book we prefer to separate between robust and reactive, as we consider that both the models and the behavior of the schedulers differ. Whereas for robust scheduling the model remains unchanged throughout the execution of the schedule (the probabilities do not change), for reactive scheduling the available model for the priorities of the passes updates constantly. Regarding the behavior of the schedulers, the robust scheduler is static (the schedule is calculated only once), but reactive scheduling recomputes a new schedule for every change.

In the following section we will present a simplified model for the changes in the priorities, which will be the case in practical scenarios. This model bounds the number of passes changing priority at the times $t \ : \ P(t) \neq P(t - \Delta t)$.

7.2 Restricted Reactive SRS: Single Pass Update Model

In the rest of the chapter we focus on the case where only one pass changes priority at a time. We will extend this model to cover the case where a small number of passes change priorities simultaneously. Finally we briefly consider the case where all the passes change priorities for completeness.

7.2.1 Overview of the Algorithm

The computation of the optimal schedule is divided into *preprocessing* and *recomputation*, illustrated in Fig. 7.2. For a detailed example of the algorithm see § 7.4.

Fig. 7.3 Preprocessing
phases

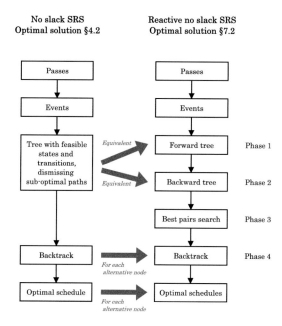

7.2.2   *Preprocessing*

We divide the preprocessing into four stages:

- *Phase 1: Forward graph creation*: basic SRS graph generation.
- *Phase 2: Backward graph creation*: similar to basic SRS graph generation, but from later to earlier times.
- *Phase 3: Best pairs in stages search*: overlay of the two previous graphs for finding alternative longest paths.
- *Phase 4: Alternative paths search*: backtracking from best pairs for providing the alternative schedule for every possible change in priority.

Note that the preprocessing will only provide alternative schedules for changes before the start time of the next pass, so that it has to be repeated for all the passes after the next pass. We show the relations with the algorithm for basic SRS in Fig. 7.3.

7.2.2.1   Phase 1: Forward Graph Creation

We present a variant of the algorithm introduced in § 4.2, in which we make the frontiers coincide with the stages. The generation of events and initialization is exactly as in § 4.2. We now detail the creation of start and end time stages.

7.2.2.1.1 Start Time Stages

We extend (4.2.5) into (7.2.1) for duplicating every node to the new stage:

$$\forall n_j \in B_{i-1} : s(n_j, g(e_i)) = 0,$$
$$s(n_j, g') \neq s(e_i) \; \forall g',$$
$$\exists^* v' = (n_j, n_x, w_x), \; n_x \in B_{i-2},$$

if $\phi(e_i) > 0$, then:

$$\exists^* n_l \in Z_i : s(n_l, g(e_i)) \triangleq s(e_i), \qquad\qquad\qquad (7.2.1)$$
$$s(n_l, g') \triangleq s(n_j, g') \; \forall g' : g' \neq g(e_i),$$
$$\exists^* v \triangleq (n_l, n_j, w_x + w(e_i)),$$
$$\exists^* n_m \in Z_i : n_m \triangleq n_j,$$
$$\exists^* v \triangleq (n_m, n_j, w_x).$$

Since each start time stage will have duplicates of those nodes in the previous stage, the frontier will coincide with the stage, so that we use (7.2.2) instead of (4.2.6):

$$B_i \triangleq Z_i. \qquad\qquad\qquad (7.2.2)$$

7.2.2.1.2 End Time Stages

For the end time stages we use (7.2.3) instead of (4.2.7), which duplicates every node not examined for deletion to the new stage:

$$\forall n_j \in B_{i-1} : s(n_j, g(e_i)) = s(e_i),$$
$$\exists^* v' = (n_j, n_x, w_j), \; n_x \in B_{i-2},$$

and if $\phi(e_i) < 0$, then also:

$$\exists^* n_y \in B_{i-1} : s(n_y, g(e_i)) = 0,$$
$$s(n_y, g') = s(n_j, g'), \; \forall g' : g' \neq g(e_i),$$
$$\exists^* v'' = (n_y, n_z, w_y), \; n_z \in B_{i-2},$$

hence: $\qquad\qquad\qquad\qquad\qquad\qquad\qquad\qquad (7.2.3)$

$$\exists^* n_l \in Z_i : n_l \triangleq n_y,$$

$$\exists^* v \triangleq \begin{cases} (n_l, n_j, w_j), & \text{if } w_j \geq w_y, \\ (n_l, n_y, w_y), & \text{if } w_j < w_y. \end{cases}$$
$$\forall n_j \in B_{i-1} : s(n_j, g(e_i)) \neq s(e_i),$$
$$\exists^* n_m \in Z_i : n_m \triangleq n_j,$$
$$\exists^* v \triangleq (n_m, n_j, w_j).$$

Again the frontier coincides with the stage, so that we use (7.2.4) instead of (4.2.8):

$$B_i \triangleq Z_i. \tag{7.2.4}$$

Let A be the algorithm in § 4.2, let A_f be the modified version of A with (7.2.1) and (7.2.2) for start time stages and (7.2.3) and (7.2.4) for the end time stages, and let G_f be the generated graph.

Corollary 7.1. *The algorithms A and A_f have the same computational complexity, and both provide the same solution.*

Proof. The two algorithms have the same criterion for creating new nodes (feasible transitions), and both propagate weights in the same way (at start time stages). Then A_f also runs in $O(N(k_1 + 1)^{k_2})$, where k_1 and k_2 are the number of ground stations and satellites, respectively.

It is easy to see that if we collapsed all the duplicated nodes towards earlier stages in G_f, we would obtain G, and therefore the longest path will be the same for both. □

7.2.2.2 Phase 2: Backward Graph Creation

Now we consider a modified version of A_f, which generates the graph backwards, that is, from later to earlier times. We denote this algorithm A_b. In this case the list of events will be examined by decreasing time order, so that we present the end time stages first. We also change the notation for the edges to b, as they have different orientations compared to those of the forward graph v. For the initialization we have $B_{2N+1} = \emptyset$, and we will use the same numeration for the nodes as that followed in Phase 1 (§ 7.2.2.1).

7.2.2.2.1 End Time Stages

End time stages for A_b are generated in the same way start time stages for A_f. Note that we keep the same numeration for stages and frontiers, as we aim at overlaying the two generated graphs later.

$$\forall n_j \in B_i : s(n_j, g(e_i)) = 0, \ s(n_j, g') \neq s(e_i) \ \forall g',$$
$$\exists^* b' = (n_j, n_x, w_x), \ n_x \in B_{i+1},$$
if $\phi(e_i) < 0$, then:
$$\exists^* n_l \in Z_{i-1} : s(n_l, g(e_i)) \triangleq s(e_i), \tag{7.2.5}$$
$$s(n_l, g') \triangleq s(n_j, g') \ \forall g' : g' \neq g(e_i),$$
$$\exists^* b \triangleq (n_l, n_j, w_x).$$

And we also duplicate all the nodes as we did in (7.2.1):

$$\forall n_j \in B_i,$$
$$\exists^* b' = (n_j, n_x, w_x),\ n_x \in B_{i+1},$$
$$\text{if }\ \phi(e_i) < 0,\ \text{then:} \tag{7.2.6}$$
$$\exists^* n_m \in Z_{i-1} : n_m \overset{\triangle}{=} n_j,$$
$$\exists^* b \overset{\triangle}{=} (n_m, n_j, w_x).$$

The new frontier coincides with the new stage:

$$B_{i-1} \overset{\triangle}{=} Z_{i-1}. \tag{7.2.7}$$

7.2.2.2.2 Start Time Stages

Consequently, start time stages for A_b are generated in the same way end time stages for A_f. Also note the numeration of stages selected to match that of the forward graph.

$$\forall n_j \in B_i : s(n_j, g(e_i)) = s(e_i),$$
$$\exists^* b' = (n_j, n_x, w_j),\ n_x \in B_{i+1},$$
and if $\phi(e_i) < 0$, then also:
$$\exists^* n_y \in B_i : s(n_y, g(e_i)) = 0,$$
$$s(n_y, g') = s(n_j, g'),\ \forall g' : g' \neq g(e_i),$$
$$\exists^* b'' = (n_y, n_z, w_y),\ n_z \in B_{i+1},$$
hence: $\tag{7.2.8}$
$$\exists^* n_l \in Z_{i-1} : n_l \overset{\triangle}{=} n_y,$$
$$\exists^* b \overset{\triangle}{=} \begin{cases} (n_l, n_j, w_j + w(e_i)), & \text{if } w_j + w(e_i) \geq w_y, \\ (n_l, n_y, w_y), & \text{if } w_j < w_y, \end{cases}$$
$$\forall n_j \in B_i : s(n_j, g(e_i)) \neq s(e_i),$$
$$\exists^* n_m \in Z_{i-1} : n_m \overset{\triangle}{=} n_j,$$
$$\exists^* b \overset{\triangle}{=} (n_m, n_j, w_j).$$

Again the frontier will coincide with the stage:

$$B_{i-1} \overset{\triangle}{=} Z_{i-1}. \tag{7.2.9}$$

Note that this modified algorithm sums the priorities just before the selection of the highest cumulated priority path, and the generated graph is offset so that the nodes coincide with those of the graph generated by A_f.

Let A_b be the modified version of A_f with (7.2.5), (7.2.6), and (7.2.7) for end time stages and (7.2.8) and (7.2.9) for the start time stages, and examining the events ordered by decreasing time.

Corollary 7.2. *The algorithms A_f and A_b have the same computational complexity, and both provide the same solution.*

Proof. Algorithm A_b can be obtained by inverting the ordering of all the start and end times of all the passes, that is: $t'_{s_l} = t_0 + T - t_{s_l}$ and $t'_{e_l} = t_0 + T - t_{e_l}$ and applying the algorithm A_f. We only change the stages at which the priorities associated to the passes are added to the cumulated metric (7.2.8) to prepare the overlay of the graphs. Therefore the two algorithms have the same computational complexity.

Let G_f be the graph generated by the algorithm A_f and G_b be the one generated by the algorithm A_b. It is easy to see that the edges indicate the longest path from every node to the terminal node of the graphs, in the case of G_f the node $n_0 \in Z_0$, and for G_b the last created node, in the last stage Z_{2N}.

Since both algorithms calculate the longest paths from all the nodes to the first node (n_0 for A_f, and the last node for A_b), and given that both algorithms follow the same criterion when selecting paths with the same priority [(7.2.3) for A_f and (7.2.8) for A_b], the longest path calculated by the two algorithms is the same.   □

7.2.2.3   Phase 3: Best Pairs in Stages Search

In this phase we overlay the graphs G_f and G_b, and we show that given any node (state of the system) we can easily find the best schedule that includes this node, and also calculate its associated metric.

Lemma 7.3. *Let $w(v_i)$ be the weight associated to the edge in G_f which origin is n_i, and $w(b_i)$ the weight associated to the edge in G_b which origin is also n_i. The weight $w(n_i) = w(v_i) + w(b_i)$ associated to the node n_i is the metric of the best path from n_0 to the last node $n_{end} \in Z_{2N}$ that includes the node n_{end}.*

Proof. Let $L_f(n_0, n_i)$ be the set of nodes resulting from backtracking G_f from n_i to n_0, and let $L_b(n_i, n_{end})$ be the set of nodes resulting from backtracking G_b from n_i to the last node $n_{end} \in Z_{2N}$.

From the definition of the algorithms A_f and A_b, the graphs G_f and G_b have the same nodes (all the possible states of the system). Given that $L_f(n_0, n_i)$ is the longest path from n_i to n_0, and $L_b(n_i, n_{end})$ from n_i to the last node of the graph, we have that $L_f(n_0, n_i) \cup L_b(n_i, n_{end})$ is the longest path from n_0 to n_{end} that includes node n_i. Given that we are increasing the cumulated metrics in the weights of the edges only in start time stages in A_f and A_b [(7.2.1) for $w(v_i)$ and (7.2.8) for $w(b_i)$], it is easy to see that $w(n_i) = w(v_i) + w(b_i)$.   □

We now define the best pair of nodes for each stage:

Definition 22. Let Z^+ be the subset of start times stages, and let e_i be the event associated to the start time stage Z_i^+. Let the current state of the system be associated to a node in Z_k. For every stage $Z_i^+ \in Z^+$ we define the best pair of nodes $R_i(k) = \{n^*(i, k), n^0(i, k)\}$ as the ones with maximal associated weight, one of them in which the pass associated to this stage is tracked, and the other one in which the pass is not tracked:

$$n^*(i, k) = \arg \max(w(n_l)) \; \forall n_l \in Z_i^+ : s(n_l, g(e_i)) = s(e_i), \qquad (7.2.10)$$

$$n^0(i, k) = \arg \max(w(n_l)) \; \forall n_l \in Z_i^+ : s(n_l, g(e_i)) = 0. \qquad (7.2.11)$$

Let A_p be the algorithm that finds the best pairs of nodes for all the start time stages.

7.2.2.4 Phase 4: Alternative Paths Search

Let $L = L_\mathrm{f}(n_0, n_\mathrm{end}) = L_\mathrm{b}(n_0, n_\mathrm{end})$ be the longest path. It is easy to see that L will include, for each stage, one of the members of the best pair of nodes associated to that stage. Let $L(n_i) = L_\mathrm{f}(n_0, n_i) \cup L_\mathrm{b}(n_i, n_\mathrm{end})$ be the longest path that includes the node n_i. For every start time stage we recompute the longest path that includes the smaller weight element of the best pair, that is, $L(n^*(i, k))$ if $n^*(i, k) \notin L$, and $L(n^0(i, k))$ if $n^*(i, k) \in L$.

Let A_r denote the algorithm for backtracking all the alternative longest paths. Since there are N start time stages, the recomputation (backtracking) of all the alternative longest paths takes $O(N^2)$.

Corollary 7.3. *The preprocessing resulting from applying the forward graph algorithm (A_f), the backward graph algorithm (A_b), the best pairs search algorithms (A_p), and the alternative longest paths algorithm (A_r) runs in $O(N^3 + N^2(k_1 + 1)^{k_2})$.*

Proof. The selection algorithm A_p will for the worst case take as much as the generation of the graphs G_f and G_b through A_f and A_b, respectively, which takes $O(N(k_1 + 1)^{k_2})$. Finally A_r takes $O(N^2)$, so that the preprocessing at the current state takes $O(N^2 + N(k_1 + 1)^{k_2})$.

Finally, note that along execution, every pass tracked or dismissed will require an update of the graph, and recomputation of the best pairs. Therefore we need to repeat the preprocessing for every pass tracked. This would mean to go to the next node that changes state in G_b, recompute G_f (basically prune edges and propagate weights), and recalculate the best pairs, repeating this process for all the start time stages. Since there are N start time stages, repeating the preprocessing for each node of the longest path that belongs to a start time stage will take $O(N^3 + N^2(k_1 + 1)^{k_2})$. □

Let $W_i(k) = \{w(n^*(i, k)), w(n^0(i, k))\}$ be the weights associated to the best pair $R_i(k)$. Then the output of the preprocessing at a current state in Z_k will be:

$$
\begin{array}{c}
R_i(k'), \; W_i(k'), \; P(L(n^*(i, k'))), \; P(L(n^0(i, k'))), \\
\forall i, k' \; : \; i > k' \geq k \wedge Z_i, Z_{k'} \in Z^+.
\end{array}
\qquad (7.2.12)
$$

The main ideas of this preprocessing are that *(a)* if Z_x is the start time stage associated to the pass that changes priority, after the change in priority the graph G_f remains unchanged from stage Z_0 to Z_x, and the graph G_b from Z_x to Z_{2N}; and *(b)* the nodes in Z_x that have that pass tracked will change priorities, but *the elements in the best pair of nodes will not change.* This is because all the nodes where the pass is tracked add the same priority change (so that $n^*(i,k)$ remains unchanged), and the nodes where the pass is not tracked do not change priorities (so that $n^0(i,k)$ also remains unchanged).

In summary, the presented preprocessing finds all the optimal schedules at all the possible stages of execution and for all the possible single variations of priority, and provides the thresholds for the priorities at which the optimal schedule has to be replaced by the alternative provided schedule.

7.2.3 Recomputation

In this section we provide the threshold values for the priorities of the passes.

Corollary 7.4. *The longest path after a pass changes priority can be computed in $O(1)$ after the preprocessing.*

Proof. Let a pass p_z (associated to the start time stage Z_i) change its priority at time t_c from w_z to w_z'. Let $\Delta w_z = w_z' - w_z$. Let Z_k be the start time stage with smallest associated event time t_k still greater than t_c, that is $\nexists Z_l : t_c < t_l < t_k$. Let $w'(n^*(i,k))$ and $w'(n^0(i,k))$ be the new weights associated to these nodes:

$$w'(n^*(i,k)) = w(n^*(i,k)) + \Delta w_z, \qquad (7.2.13)$$

$$w'(n^0(i,k)) = w(n^0(i,k)). \qquad (7.2.14)$$

Let L and L' be the longest paths (highest metric) before and after the priority change, respectively. Then:

$$L' = \begin{cases} L(n^*(i,k)) & \text{if } w'(n^*(i,k)) \geq w'(n^0(i,k)), \\ L(n^0(i,k)) & \text{if } w'(n^*(i,k)) < w'(n^0(i,k)). \end{cases} \qquad (7.2.15)$$

And therefore the updated optimal schedule is:

$$P^*(t_c) = P^*(t_c - \Delta t)[t_0, t_c] \cup \begin{cases} P(L(n^*(i,k))) & \text{if } w'(n^*(i,k)) \geq w'(n^0(i,k)), \\ P(L(n^0(i,k))) & \text{if } w'(n^*(i,k)) < w'(n^0(i,k)). \end{cases}$$

$$(7.2.16)$$

Since the check only involves the comparison of the new weights of the best pair, the computation takes $O(1)$.   □

We can also calculate the threshold value $w_{th}^k(p_z)$ for the pass p_z (associated to the start time stage Z_i) and at the current stage Z_k. From (7.2.13)–(7.2.15):

$$w_{th}^k(p_z) = \max\{0, \min\{1, w(p_z) + w(n^0(i,k)) - w(n^*(i,k))\}\}. \qquad (7.2.17)$$

That is, if $w(p_z) < w_{th}^k(p_z) < w'(p_z)$, or if $w(p_z) > w_{th}^k(p_z) > w'(p_z)$, then the alternative path has to be selected.

7.3   Restricted Reactive SRS: Multiple Pass Update Model

If the model of the problem includes several passes changing priority at the same time it will not be possible to forecast alternative schedules without involving heavy computations. We provide a modification of the process presented in § 7.2 for reducing the computation time for the optimal schedule:

- *Preprocessing.*

    - *Phase 1: Forward graph creation.* Equivalent to that presented in § 7.2.2.1.
    - *Phase 2: Backward graph creation.* Equivalent to that presented in § 7.2.2.2.

- *Recomputation.*

    - *Phase 3: Backward graph repair.* Let p_a be the pass with earliest start time that changes priority, and p_z that with the latest start time. The *Phase 3* will recompute the graph G_b from Z_i (start time stage associated to the pass p_z) to Z_y (start time stage associated to p_a), following algorithm A_b. Let $M_c = z - a$. Then this recomputation will take $O(M_c(k_1 + 1)^{k_2})$.
    - *Phase 4: Best element in stage search.* At this point the priority change is already known, so that it is only necessary to find the best element of the stage Z_y. Since there will be $(k_1 + 1)^{k_2}$ elements in a stage for the worst case, the search of the best element will take $O((k_1 + 1)^2)$.
    - *Phase 5: Optimal path recomputation.* The final step is backtracking from the node found in the previous phase to the two extremes of the graphs G_f and G_b, which takes $O(N)$.

The preprocessing will therefore take $O(N^2(k_1 + 1)^{k_2})$, and the recomputation $O(M_c(k_1 + 1)^{k_2})$ (we assume that $O(M_c(k_1 + 1)^{k_2}) > O(N)$).

It can be noted that for the worst case in the number of changes, that is, $M_c = N$, the preprocessing does not provide any benefit as the complete graph has to be recomputed through the algorithm presented in § 4.2.

7.4   Schedule Computation Example

We show a simple example for the calculation of the alternative schedules. We consider the restricted problem presented in § 7.2, where only one pass may change priority at a time.

Fig. 7.4 Set of passes

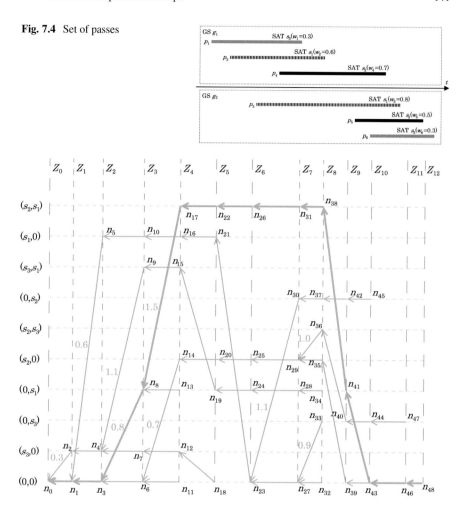

Fig. 7.5 Phase 1: forward graph generation

The set of passes $P(t) = \{p_1, p_2, \ldots, p_6\}$ with associated priorities $w_1 = 0.3$, $w_2 = 0.6$, $w_3 = 0.8$, $w_4 = 0.7$, $w_5 = 0.5$, and $w_6 = 0.3$ at $t = t_0$ is represented in Fig. 7.4. The optimal set of passes for $P(t_0)$ is $P^*(t_0) = \{p_3, p_4\}$, with $\|P^*(t_0)\|_{\Sigma w} = 1.5$. We will show the preprocessing for the stage Z_1.

Phase 1: Forward Graph Creation

We show the forward graph G_f in Fig. 7.5. Note that compared to the graph from § 4.5, besides the additional passes, in this case nodes are duplicated through following stages if they remain active.

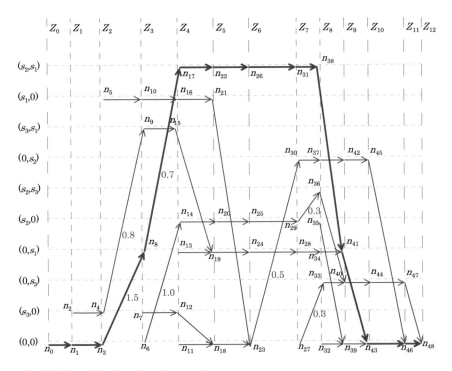

Fig. 7.6 Phase 2: backward graph generation

Phase 2: Backward Graph Creation

We show the backward graph G_b in Fig. 7.6. It can be seen that the sets of nodes of G_b and G_f coincide, only the edges change.

Phase 3: Best Pairs in Stages Search

We show in Fig. 7.7 the overlay of the graphs G_f and G_b, and the best pairs $R_i(1)$ for each stage i. We enumerate these pairs and their associated weights as follows:

$$
\begin{aligned}
p_1 \to Z_1 \to R_1(1) &= \begin{cases} n^*(1,1) = n_2 &: w(n_2) = 1.1, \\ n^0(1,1) = n_1 &: w(n_1) = 1.5, \end{cases} \\
p_2 \to Z_2 \to R_2(1) &= \begin{cases} n^*(2,1) = n_5 &: w(n_5) = 1.1, \\ n^0(2,1) = n_3 &: w(n_3) = 1.5, \end{cases}
\end{aligned}
\tag{7.4.1}
$$

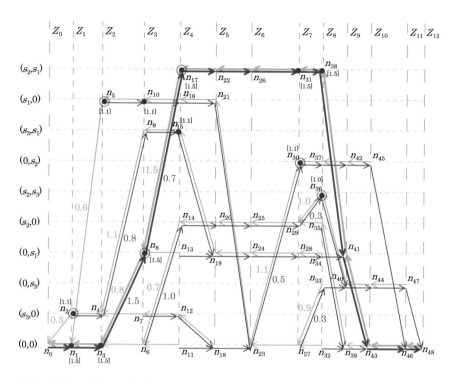

Fig. 7.7 Phase 3: best pairs in stages search

$$p_3 \rightarrow Z_3 \rightarrow R_3(1) = \begin{cases} n^*(3,1) = n_8 & : w(n_8) = 1.5, \\ n^0(3,1) = n_{10} & : w(n_{10}) = 1.1, \end{cases}$$

$$p_4 \rightarrow Z_4 \rightarrow R_4(1) = \begin{cases} n^*(4,1) = n_{17} & : w(n_{17}) = 1.5, \\ n^0(4,1) = n_{15} & : w(n_{15}) = 1.1, \end{cases}$$

$$p_5 \rightarrow Z_7 \rightarrow R_7(1) = \begin{cases} n^*(7,1) = n_{30} & : w(n_{30}) = 1.1, \\ n^0(7,1) = n_{31} & : w(n_{31}) = 1.5, \end{cases}$$

$$p_6 \rightarrow Z_8 \rightarrow R_8(1) = \begin{cases} n^*(8,1) = n_{36} & : w(n_{36}) = 1.0, \\ n^0(8,1) = n_{38} & : w(n_{38}) = 1.5. \end{cases}$$

Phase 4: Alternative Paths Search

We now show the alternative paths for every stage before pass p_1 is executed. We indicate the thresholds for the new priority values calculated as in (7.2.17).

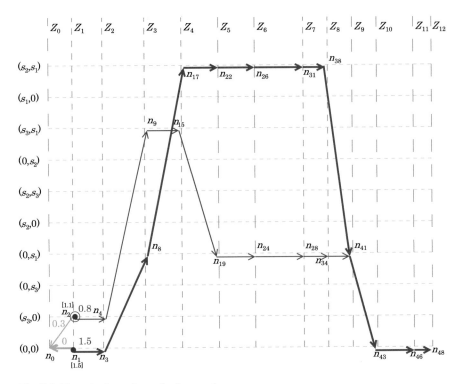

Fig. 7.8 Phase 4: alternative paths for pass 1

$$R_1(1) \ : \ P^{*\prime} = \begin{cases} P(L'(n_2)) = \{p_1, p_3\} & : w_1' > 0.7, \\ P(L'(n_1)) = \{p_3, p_4\} & : w_1' \leqslant 0.7, \end{cases}$$

$$R_2(1) \ : \ P^{*\prime} = \begin{cases} P(L'(n_5)) = \{p_2, p_5\} & : w_2' > 1, \\ P(L'(n_3)) = \{p_3, p_4\} & : w_2' \leqslant 1, \end{cases}$$

$$R_3(1) \ : \ P^{*\prime} = \begin{cases} P(L'(n_8)) = \{p_3, p_4\} & : w_3' \geqslant 0.4, \\ P(L'(n_{10})) = \{p_2, p_5\} & : w_3' < 0.4, \end{cases}$$

$$R_4(1) \ : \ P^{*\prime} = \begin{cases} P(L'(n_{17})) = \{p_3, p_4\} & : w_4' \geqslant 0.3, \\ P(L'(n_{15})) = \{p_2, p_5\} & : w_4' < 0.3, \end{cases} \tag{7.4.2}$$

$$R_7(1) \ : \ P^{*\prime} = \begin{cases} P(L'(n_{30})) = \{p_2, p_5\} & : w_5' > 0.9, \\ P(L'(n_{31})) = \{p_3, p_4\} & : w_5' \leqslant 0.9, \end{cases}$$

$$R_8(1) \ : \ P^{*\prime} = \begin{cases} P(L'(n_{36})) = \{p_4, p_6\} & : w_6' > 0.8, \\ P(L'(n_{38})) = \{p_3, p_4\} & : w_6' \leqslant 0.8. \end{cases}$$

Note that the schedules associated to the nodes n_1, n_3, n_8, n_{17}, n_{31}, and n_{38} are equal to the optimal schedule $P^*(t_0)$. Also, given that those schedules associated to n_5, n_{10}, n_{15}, and n_{30} coincide ($\{p_2, p_5\}$), we show only the alternative paths for n_2 (pass 1, in Fig. 7.8), n_5 (pass 2, in Fig. 7.9), and n_{36} (pass 6, in Fig. 7.10).

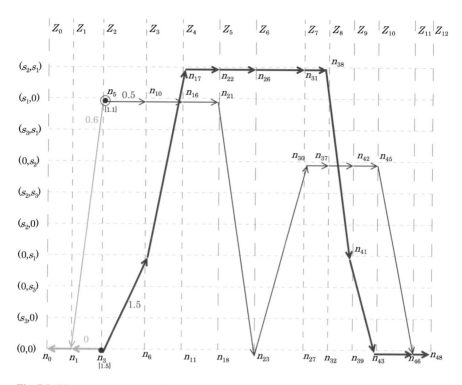

Fig. 7.9 Phase 4: alternative paths for pass 2

Completing the preprocessing would require to compute the rest of best pairs and associated paths as indicated in (7.2.12).

Phase 5: Recomputation

Let us suppose that there is a change in priority at $t < t_s(p_1)$. Then it would only take to check (7.4.2) for obtaining the alternative schedule.

7.5 Summary

We extend the notation in Chap. 3 (§ 3.3) to cover the reactive SRS problem. For that purpose we modify the symbol representing the priority to show the time dependency with uncertain change times: $\mathfrak{w}(\tilde{t})_j$.

We summarize the results presented in this chapter into Table 7.1 for easy consultation, and we add the single scheduling entity cases for completeness, for

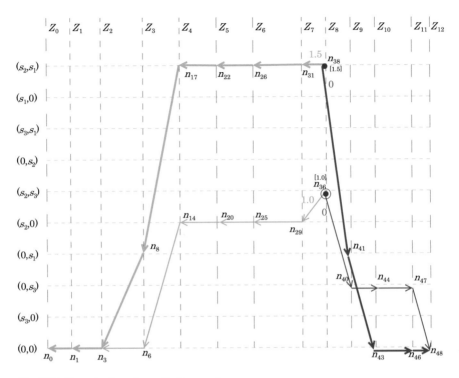

Fig. 7.10 Phase 4: alternative paths for pass 6

Table 7.1 Optimal solutions for Reactive SRS

Problem	Preprocessing	Recomputation	References Approx.	Optimal
$1\,\|\,\tau_j,\overline{\overline{p}}_{ij},C_\Sigma\,\|\,\sum \mathfrak{w}(\tilde{t})_j\mathfrak{U}_j$ $(M_c=1)$	$O(N^3)$	$O(1)$	–	§ 7.2
$1\,\|\,\tau_j,\overline{\overline{p}}_{ij},C_\Sigma\,\|\,\sum \mathfrak{w}(\tilde{t})_j\mathfrak{U}_j$ $(1<M_c<N)$	$O(N^2 k_1)$	$O(M_c k_1)$	–	§ 7.3
$1\,\|\,\tau_j,\overline{\overline{p}}_{ij},C_\Sigma\,\|\,\sum \mathfrak{w}(\tilde{t})_j\mathfrak{U}_j$ $(M_c=N)$	–	$O(N)$	–	§ 6.5
$\Re\,\|\,\tau_j,\overline{\overline{p}}_{ij},C_\Sigma\,\|\,\sum \mathfrak{w}(\tilde{t})_j\mathfrak{U}_j$ $(M_c=1)$	$O(N^3 + N^2(k_1 + 1)^{k_2})$	$O(1)$	–	§ 7.2
$\Re\,\|\,\tau_j,\overline{\overline{p}}_{ij},C_\Sigma\,\|\,\sum \mathfrak{w}(\tilde{t})_j\mathfrak{U}_j$ $(1<M_c<N)$	$O(N^2(k_1 + 1)^{k_2})$	$O(M_c(k_1 + 1)^{k_2})$	–	§ 7.3
$\Re\,\|\,\tau_j,\overline{\overline{p}}_{ij},C_\Sigma\,\|\,\sum \mathfrak{w}(\tilde{t})_j\mathfrak{U}_j$ $(M_c=N)$	–	$O(N(k_1 + 1)^{k_2})$	–	§ 4.2

which $k_2 = 1$, and given that in general $N > k_1$, the complexity of the preprocessing for $1 \mid \tau_j, \overline{\overline{p}}_{ij}, C_\Sigma \mid \sum \mathfrak{w}(\tilde{t})_j\mathfrak{U}_j$ with $M_c = 1$ simplifies to $O(N^3)$. Also, for the problem $1 \mid \tau_j, \overline{\overline{p}}_{ij}, C_\Sigma \mid \sum \mathfrak{w}(\tilde{t})_j\mathfrak{U}_j$ with $M_c = N$, we include the reference to the algorithm presented in § 6.5. For the three considered single scheduling entity cases we also assume that $k_1 \gg 1$. Note that all the problems include multiple resources, priorities, no-preemption and no-slack, and consider a fixed number of scheduling entities.

We show the relations among the presented problems in Fig. 7.11.

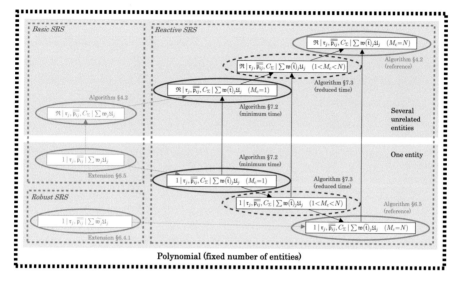

Fig. 7.11 Complexity of reactive SRS and relations with basic SRS and robust SRS

Acknowledgements This research was performed while the author held a National Research Council Research Associateship Award at the Air Force Research Laboratory (AFRL).

References

1. J.C. Pemberton, L.G. Greenwald, On the need for dynamic scheduling of imaging satellites, in *FIEOS 2002 Conference, Denver* (2002)
2. D. Ouelhadj, S. Petrovic, A survey of dynamic scheduling in manufacturing systems. J. Sched. **12**(4), 417–431 (2009)

Chapter 8
Summary

In this chapter we present a summary of all the problems solved in the book, as well as their relations, and we also provide some lines of inquiry for future work.

8.1 Conclusions

In this book we have solved a number of problems in the SRS domain which were only approached through suboptimal solution algorithms by existing literature. At the time of writing, to the best of our knowledge, only optimal solutions were known for the basic SRS problems with no priorities and fixed times requests, or equivalently $1 \mid r_j, \overline{\overline{p_{ij}}}, C_\Sigma \mid \sum \mathfrak{U}_j$ and $\mathfrak{P} \mid r_j, \overline{\overline{p_{ij}}}, C_\Sigma \mid \sum \mathfrak{U}_j$ (§ 3.3), which are subproblems of those for which we have presented optimal solutions.

We show in Fig. 8.1 all the problems solved in this book, presented in a similar way to the summaries of previous chapters (§ 3.5, § 4.7, § 5.8, and § 7.5). For those problems for which we did not provide the optimal solution, we provided proof of intractability, even under practical conditions: fixed number of entities (FNE) and discretized time (Δt). Besides the relations for the problems with reduced information for noncooperative SRS, we also include the relations provided in previous chapters relating dynamic, robust, and noncooperative SRS, with dashed arrows pointing towards the problems with reduced information.

8.2 Future Work

We have only presented the SRS problem and its main variants. Some applications would require combinations of these problems, like distributed scenarios involving

© Springer International Publishing Switzerland 2015
A.J. Vázquez Álvarez, R.S. Erwin, *An Introduction to Optimal Satellite Range Scheduling*, Springer Optimization and Its Applications 106,
DOI 10.1007/978-3-319-25409-8_8

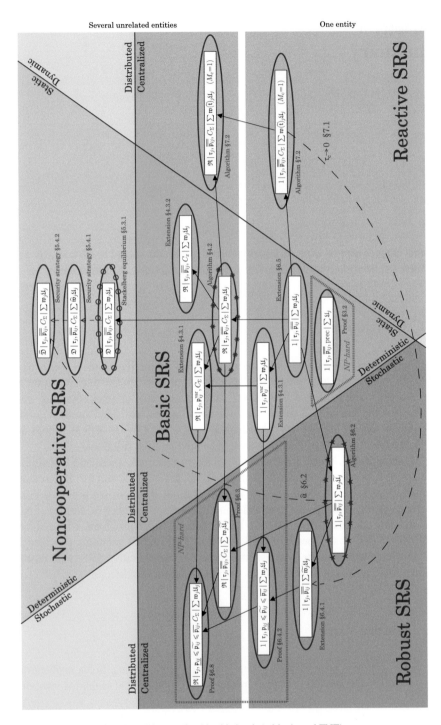

Fig. 8.1 Summary of SRS problems solved in this book (with Δt and FNE)

uncertainty (*robust noncooperative SRS*) or updates on the requests (*reactive noncooperative SRS*), or even changing uncertainty models (*robust reactive SRS*). Based on our previous results, given that robust SRS is NP-hard for multiple entities, we conjecture that robust noncooperative SRS is NP-hard. Following the same reasoning we also conjecture that robust reactive SRS is NP-hard for multiple entities.

More specific lines of future work have been presented for the *noncooperative SRS* problem. In Proposition 5.1 we have considered that for strategically equivalent games with the same payoff for the leader, the players deterministically select the path in which the differing pass is tracked (5.3.11). Thus, a change in this case (i.e., selecting n_j instead of n_y) would only change the payoffs of the rest of the players (not necessarily decreasing them), which realize the change (perfect information), and thus are able to make decisions in this sense too in the future (altruistic behavior or retaliation). This iterated game overlying on the SRS Game with perfect information is thus a consequence of the degree of freedom on the selection of same payoff strategically equivalent games. Also as introduced in § 5.8, versions of the game with no information on the history of play are left as future work. We have also introduced lines of future work in the migration of these results to machine scheduling, and in the introduction of cooperative game theory in § 5.5.

For *reactive SRS* we have provided the optimal solution for a single pass changing priority, future work should include improving the results provided in § 7.3 for multiple passes changing priority. Other variants could include late notice of the changes in priority, which would be the base for tackling limited information versions of the reactive noncooperative SRS problem.

Acknowledgements This research was performed while the author held a National Research Council Research Associateship Award at the Air Force Research Laboratory (AFRL).

Glossary

Algorithm Stepwise procedure for finding the solution of a problem.

Backtracking Traveling a tree graph from a leaf node towards the root node.

Backward graph Directed acyclic graph modeling the optimal transitions among all the possible states of the system, with root on the end time of the latest end time pass, and with edges pointing from earlier to later nodes.

Belief network Directed acyclic graph which nodes represent random variables, and edges conditional dependencies among them.

Best pair of nodes in stage Nodes in a stage of the directed acyclic graph modeling the optimal transitions among all the possible states of the system, one of them including the pass associated with that stage and the other not, and only one of them belonging to the longest path of the graph depending on the priority of that pass.

Class NP (Nondeterministic polynomial). Set of problems which solution can be verified with a polynomial time algorithm.

Class P (Polynomial). Set of problems for which a polynomial time algorithm is known.

Clique problem Problem of finding a subgraph with a certain number of nodes, with all pairs of nodes connected.

Complexity We refer to the computational complexity of an algorithm as the time required to obtain a solution as a function of the size of the input of the problem.

Conflict Two passes are said to be conflicting if they are both time-overlapping and associated with the same satellite or ground station, or if they are associated with the same request.

Decision problem Problem which solution is a boolean value.

Directed acyclic graph (DAG) Graph with nodes connected by directed edges, and which does not contain any directed path starting and ending in the same node.

Distributed scheduler Scheduler based on the decisions of multiple agents with different objectives and/or available information.

© Springer International Publishing Switzerland 2015
A.J. Vázquez Álvarez, R.S. Erwin, *An Introduction to Optimal Satellite Range Scheduling*, Springer Optimization and Its Applications 106,
DOI 10.1007/978-3-319-25409-8

Due time of a request Deadline of the request.

Earth observation Aerospace application field which objective is the imaging of specific spots of the Earth, generally performed through low Earth orbit (LEO) satellites.

Event Tuple associated with start and end times of the passes used for the generation of the graph modeling the transitions among all the possible states of the system.

Executed schedule Set of passes that have been tracked, that is, the communication (or observation) has been successful, at the end of the scheduling horizon.

Expected metric In a scenario involving uncertainty, the expected metric of the schedule provides the average metric if the schedule were executed a large number of times.

Extensive form representation Tree graph representing all the possible actions of the players in a game, with each level of the graph representing the action of a player, and with the leaves of the tree having associated payoffs for all the players.

Feasible schedule Schedule with no conflicts.

Fixed interval This term is equivalent to no-slack.

Fixed number of entities (FNE) The number of ground stations and satellites is constant.

Follower In a feedback Stackelberg game, player that takes action after the leader.

Forward graph Directed acyclic graph modeling the optimal transitions among all the possible states of the system, with root on the start time of the earliest start time pass, and with edges pointing from later to earlier nodes.

Frontier Set of nodes that are checked in the creation of the next stage for the graph modeling the transitions among all the possible states of the system.

Game theory Discipline that models multi-agent strategic decision making.

General scheduling Also called machine scheduling, it is a field of research focused on the allocation of jobs in a set of resources.

Geostationary orbit Circular equatorial prograde orbit with the same orbital period as the Earth, so the satellite remains over the same spot on Earth at all times.

Graph Set of nodes connected by edges.

Greedy algorithm Algorithm which iteratively takes locally optimal solutions according to a heuristic, but in general does not guarantee finding the optimal global solution.

Ground segment Facilities for the communication, command and operation of the satellite, and for reception and distribution of data in a satellite mission. It is composed by one or various ground stations, mission control centers, data distribution centers, and an interconnection network.

Ground station Telecommunication facilities for communicating to a satellite.

Ground station network Set of ground stations grouped for increasing the coverage and capabilities of the satellite mission.

Heuristic Rule followed in a suboptimal solution algorithm, like the ordering of tasks in a greedy algorithm.

Heuristic algorithm Algorithm that does not guarantee finding the optimal solution.

Indegree of a node Number of edges ending in the node.

Interval graph Undirected graph where nodes represent intervals, and edges among pairs of nodes the intersection among pairs of intervals.

Job Input unit in a scheduling problem basically associated with a resource for a certain period of time.

Leader In a feedback Stackelberg game, player that takes action first.

Leader satellite Satellite that decides which edge is dismissed in an end time stage in noncooperative SRS.

Leaf node (of directed tree graph) Node with null indegree in a tree directed towards the root.

Line of sight (LOS) Two objects are in line of sight if the imaginary line that connects them does not intersect any object. More specifically, a satellite is in line of sight with a ground station if it is above its horizon.

Longest path For a pair of nodes in a graph, it is the path between them which sum of priorities associated with its edges is maximal.

Low Earth orbit (LEO) Earth orbit with an altitude smaller than around 10^3 km.

m-ary capacity Constraint that applies to scenarios where satellites may communicate to at most m ground stations at the same time (or vice versa).

Machine scheduling See general scheduling.

Maximin payoff Maximum payoff achievable by a player for the worst case.

Metric Sum of the priorities (preference value) of a feasible schedule.

Mission control center Facilities for the operation of the satellite.

Multiple-interval graph Undirected graph where nodes represent sets of intervals, and edges among pairs of nodes the intersection among pairs of intervals from the two corresponding sets of intervals.

Multiple Resource Range Scheduling Problem (MuRRSP) Term used in literature for Satellite Range Scheduling (SRS) for multiple ground stations and satellites.

Multiply connected belief network Belief network that contains at least a pair of nodes connected by different paths.

Nash equilibrium Solution of a game where no player can improve its payoff by unilaterally deviating from this solution.

Noncooperative game theory Branch of game theory that considers players which have conflicting interests.

NP-complete A decision problem is said to be NP-complete if its solution can be verified in polynomial time (i.e., it belongs to the class NP) and it can be transformed in polynomial time into another known NP-complete problem.

NP-hard A problem is NP-hard if it is at least as complicated as an NP-complete problem (i.e., it can be transformed into another known NP-complete problem).

Operations plan Set of rules and procedures for the operation of the satellite from which the communication (or observation) requests are generated.

Optimal schedule Feasible schedule with maximal metric.

Outdegree of a node Number of edges originating in the node.

Pass Tuple modeling an interaction with fixed start and end times between a ground station and a satellite, and with an assigned priority. It is the analogous to fixed-interval job in general scheduling.

Path (directed) Given a pair of nodes in a directed tree, a path is a directed sub-tree that has one of those nodes as root and the other node as the only leaf.

Payoff Metric specific for each player in a game.

Precedence Constraint applicable to an SRS problem if a pass requires another pass already executed to initiate.

Preemption Constraint applicable to an SRS problem if a pass can be interrupted along its execution.

Price of anarchy (PoA) Ratio between the sum of the payoffs for all players and the centralized metric of the optimal solution.

Priority (or suitability) function Function that models the preference of a request depending on its duration and location in the visibility window.

Priority of a pass Value for modeling preference relations among passes.

Probabilistic inference Calculation of probabilities in a belief network.

Rational player In game theory, player that aims at maximizing its own associated metric.

Reactive scheduler Scheduler that recomputes the optimal schedule after the priority changes are known, aiming at reducing the recomputation time.

Redundancy Constraint applicable to an SRS problem if conflicts of some kind are allowed. See m-ary capacity.

Release time of a request Time after which the interaction can start.

Request Tuple modeling an interaction between a ground station and a satellite, which must be executed between a release time and a due time, which duration is smaller or equal to the difference of these times and which priority depends on its start time and duration. It is the analogous to job in general scheduling.

Robust schedule Schedule, not necessarily feasible, which has the highest expected metric.

Robust scheduler Scheduler that generates a static schedule taking into account uncertainty.

Root node (of directed tree graph) Node with null outdegree (in a tree graph directed towards the root).

Satellite Orbiting spacecraft with communication, observation, or experimentation applications.

Satellite communication We refer to communication involving a satellite and a ground station.

Satellite mission Project for launching and operating a satellite with a particular objective.

Satellite operator Staff in charge of planning, scheduling and commanding the satellite.

Satellite Range Scheduling (SRS) Problem of allocating a set of time intervals (requests) among two kinds of entities (generally ground stations and satellites). These intervals may have a variety of constraints, like mutual exclusion between some of them, and preference relations.

Schedule Subset of the initial set of passes.

Scheduler System which, given an initial set of requests, provides a schedule satisfying certain conditions depending on the type of problem.

Scheduling horizon Time window that contains all the time intervals associated with the initial set of requests.

Security strategy Set of actions that guarantees to the player a maximin payoff regardless of the actions of the other players.

Selfish A selfish satellite is that which aims at maximizing its own associated metric, instead of the (centralized) metric of the schedule.

Single Resource Range Scheduling Problem (SiRRSP) Term used in literature for Satellite Range Scheduling (SRS) for a single ground station or satellite.

Slack Applied to requests that have no fixed start and end times.

Social welfare (SW) In noncooperative SRS, it is equivalent to the metric of the schedule.

Stackelberg equilibrium Solution of a Stackelberg game in which no player can improve its payoff by unilaterally deviating from this solution.

Stackelberg game Game where players take actions in pre-specified turns.

Stage Set of nodes associated with the same event in a forward (or backward) graph.

Static scheduler Scheduler that provides a schedule only once before the scheduling horizon.

Suboptimal solution algorithm Algorithm that provides a solution which performance (metric of the schedule) is smaller than that of the optimal.

Task See job.

Time overlapping Two intervals are said to be time overlapping if the start time of one of them is between the start and end times of the other one.

Time varying graph (TVG) Graph which edges' presence varies along time.

Tractable We say that a problem is tractable if it can be solved in polynomial time.

Tree graph (directed towards root) Graph with a single undirected path between every pair of nodes, and with every node having unitary outdegree, except one of them (root) which has null outdegree.

Unified notation Notation widely used in general scheduling literature to classify problems.

Unitary capacity Constraint that applies to scenarios where satellites may only communicate to one ground station at the same time and vice versa.

Visibility window Time window during which there is line of sight between two objects.

Index

A
Algorithm
   noncooperative SRS
      perfect information, 83
      uncertain passes, 93
      uncertain priorities, 91
   optimal SRS
      no slack, 123
      no-slack, 51
      variable slack, 56
   reactive SRS
      alternative longest paths, 138
      backward graph, 135
      best pairs, 137
      forward graph, 133
      multiple pass update, 140
      single pass update, 132
   robust SRS
      expected metric, 112
      robust schedule, 113, 118

B
Belief network
   multiply connected, 114, 115, 122
Big O notation, 34

C
Capacity
   m-ary, 13, 30, 57
   unitary, 13, 30
Changes, 15
   dynamic, 14, 15, 130
   static, 23

Class NP, 34
Class P, 34
Classification
   noncooperative SRS, 105
   optimal SRS
      algorithms, 72
      problems, 46
   reactive SRS, 146
   robust SRS, 128
   summary, 150
Clique problem, 35
Complexity
   introduction, 34
   results, summary, 46, 104, 126, 145
Conflict, 13, 29

D
Decision problem, 34–36
Discretization step, 27

E
Entities, 25
   fixed number of, 36
Equilibrium, 93
   Nash, 93
   Stackelberg
      computation, 87
      considerations, 93
      definition, 83
Example
   calculation of priorities, 32
   conflicting passes, 30

© Springer International Publishing Switzerland 2015
A.J. Vázquez Álvarez, R.S. Erwin, *An Introduction to Optimal Satellite Range
Scheduling*, Springer Optimization and Its Applications 106,
DOI 10.1007/978-3-319-25409-8

multiply connected belief network, 115, 122
noncooperative SRS graph generation, 94
optimal SRS graph generation, 61
reactive SRS approaches, 131
reactive SRS graph generation, 140
robust SRS schedule computation, 124
scheduling process, 13
set of later non-conflicting passes, 111
transformation of request into passes, 28
types of slack, 26

F
Function
conflict indicator, 29
event generation, 52
priority (or suitability), 13, 32

G
Game
elements, 79
actions, 80
extensive form representation, 81, 99
levels, 81
players, 79
rationality, 81
feedback Stackelberg, 82
follower, 82
leader, 82, 87
information, 80
limited, 89
perfect, 82
uncertain passes, 92
uncertain priorities, 90
iterated, 151
metric
price of anarchy (PoA), 94
social welfare (SW), 94
payoff, 80
maximin, 90
vector, 85
strategically equivalent, 88
strategy, 82
theory, 77
Graph
backtracking, 54, 89
creation, 52, 85, 133, 135
end time event stage, 53, 86, 134, 135
initialization, 52, 85
start time event stage, 52, 85, 134, 136
elements
best pair of nodes, 137

edge, 52, 84
event, 51, 83
frontier, 52, 85
longest path, 54, 139
node, 52, 84
stage, 52, 84
sub-path, 84
types
backward, 135
directed acyclic (DAG), 55
forward, 133
time varying (TVG), 22, 23
tree, 54, 88
undirected, 35
Ground station, 11, 22
Ground station network, 11, 14, 21
AFSCN, 3, 16
DSN, 3, 16
ESTRACK, 3, 16

H
Heuristic
earliest deadline, 40, 59, 66, 72
earliest start time, 61, 101, 125
maximum priority, 59, 66

I
Intractable, 34, 35, 114, 151

J
Job, 37
conflict, 38
cost function, 38
due time, 38
duration, 38
optimization function, 39
precedence, 39
preemption, 39
priority / weight, 38
processing time, 37
release time, 37, 38

L
Line of sight (LOS), 11, 22
Loop cutset, 115

M
Machine, 37
general scheduling, 37, 94

parallel identical, 38, 41
single, 40
unrelated parallel, 38, 42
Mission control center, 11, 42, 78
Multiple resource range scheduling
   problems (MuRRSP), 25

N
NP-complete, definition, 34
NP-hard, definition, 34

O
Operations plan, 11, 13
Optimization problem, 34
Orbit, 11
geostationary (GEO), 13, 27
high altitude, 27
low Earth (LEO), 13, 27, 66
Outdegree, 54

P
Pass, 27
backup, 107
end time, 27
initial set, 33
number, 60
preference, 32
priority, 32
  random, 116
set of later non-conflicting, 110
start time, 27
weight, 32
Polynomial transformation, 27, 34
Precedence, 13, 31
Preemption, 13
Preprocessing, 132
Probabilistic inference, 114, 122
Probability
failure, 108
of pass to be executed, 109

R
Recomputation, 139
Recursive computation, 112, 113, 117
Redundancy, 13, 28
Request, 11, 24
constraints, 13, 24
discretized time, 27
due time, 13, 24
duration, 26

maximum, 13, 24
minimum, 13, 24
random, 119
minimum duration, 26
preference, 13
priority, 13, 24
release time, 13, 24
set of passes generated from, 27
weight, 24

S
Satellite, 11, 22
communication, 3, 11, 21
Earth observation (EOS), 3, 11, 21,
  22
operator, 11
specific model, 16
Satellite Range Scheduling (SRS)
applications, 3
definition, 3
relation with general scheduling, 94
types
  noncooperative, 77
  optimal, 49, 123
  reactive, 129
  robust, 107
Schedule, 12, 28
associated to sub-path, 84
executed, 108
feasible, 13, 32
makespan, 94
metric, 14, 32
partial results, 60, 70
robust, 109
security, 90, 92, 93
Scheduler
centralized, 50
distributed, 79
reactive, 130
robust, 109
Scheduling
fixed interval, 27
horizon, 13, 60
problem
  oversubscribed, 33
  reducibility, 39
process, 11, 12, 16
resource, 51
Selfish, 77, 101
Sensor scheduling, 3, 11, 22
Simulation
noncooperative SRS, 101
optimal SRS, 66

robust SRS, 124
scenario, 66
Single resource range scheduling problems (SiRRSP), 25
Slack, 26
  fixed, 26
  no, 26
  variable, 26
Suboptimal solution algorithm, 16, 21
  greedy, 40, 59, 61, 66, 72

T
Topology, 15
  centralized, 23, 123

complexity remarks, 60
distributed, 14, 15, 78
Transformation, 27

U
Uncertainty, 14, 15
  deterministic, 23, 78
  stochastic, 15, 107
Unified notation, 21, 37, 38, 104, 126, 145

V
Visibility window, 11, 22, 23

Printed in the United States
By Bookmasters